Nitrogen Fixation

The fixation of nitrogen – the conversion of atmospheric nitrogen to a form which plants can use – is fundamental to the productivity of the biosphere and therefore to the ability of the expanding human population to feed itself. Although the existence and importance of the process of biological nitrogen fixation have been recognized for more than a century, scientific advances over the past few decades have radically altered our understanding of its nature and mechanisms. This book provides an introductory survey of biological nitrogen fixation, covering the role of the process in the global nitrogen cycle as well as its biochemistry, physiology, genetics, ecology, general biology and prospects for its future exploitation. This new edition includes the most recent developments in the field, so providing an up-to-date and accessible account of this key biological process.

John Postgate FRS is Emeritus Professor of Microbiology at the University of Sussex and Former Director of the AFRC Unit of Nitrogen Fixation. His book publications include *The Sulphate-Reducing Bacteria*, *The Fundamentals of Nitrogen Fixation*, *Microbes and Man* and *The Outer Reaches of Life*.

Nitrogen Fixation

THIRD EDITION

John Postgate

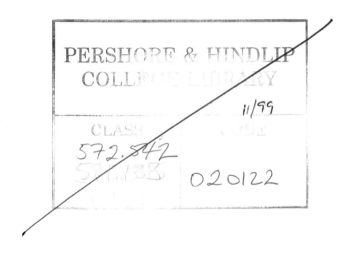

CAMBRIDGE
UNIVERSITY PRESS

PUBLISHED BY THE PRESS SYNDICATE OF THE UNIVERSITY OF CAMBRIDGE
The Pitt Building, Trumpington Street, Cambridge CB2 1RP, United Kingdom

CAMBRIDGE UNIVERSITY PRESS
The Edinburgh Building, Cambridge CB2 2RU, UK http://www.cup.cam.ac.uk
40 West 20th Street, New York, NY 10011-4211, USA http://www.cup.org
10 Stamford Road, Oakleigh, Melbourne 3166, Australia

First published 1978
Second edition 1987
Third edition 1998

Printed in the United Kingdom at the University Press, Cambridge

Typeset in Utopia 9.25/13.5pt [VN]

A catalogue record for this book is available from the British Library

Library of Congress Cataloguing in Publication data

Postgate, J. R. (John Raymond)
Nitrogen fixation/John Postgate. – 3rd ed.
       p.   cm.
Includes bibliographical references and index.
ISBN 0 521 64047 4
1. Nitrogen-fixing microorganisms.   2. Nitrogen – fixation.
3. Root-tubercles.
QR89.7.P675   1998   98–15362   CIP
572'.545–dc21

ISBN 0 521 64047 4 hardback
ISBN 0 521 64853 X paperback

Contents

Preface

The fixation of nitrogen – the conversion of atmospheric nitrogen into a form that plants can use – is a process fundamental to world agriculture. It comes about as a consequence of spontaneous, anthropogenic and biological activities. The existence and importance of the biological component have been recognized for more than a century, but scientific advances over the past few decades have radically altered our understanding of its nature and mechanisms. This book, substantially revised since the second edition, surveys biological nitrogen fixation with emphasis on recent discoveries. The place of nitrogen fixation in the global nitrogen cycle, recent developments in its biochemistry, physiology, genetics and general biology, plus future prospects for its exploitation, are all covered in an elementary manner suitable for sixth formers and first-year undergraduates.

1998   John Postgate

1   The nitrogen cycle

1.1   Biological cycles

Living things, on this planet, continually recycle the chemical elements
of which they are composed. The most familiar, day-to-day, example of
such a cycling process is the carbon cycle. In this cycle, in its simplest
form, green plants use photosynthesis to convert CO_2 from the atmosphere into plant material, while animals, by consuming the plants,
digesting them and using their carbon compounds for respiration, return the carbon atoms to the air as CO_2. Thus the carbon atoms of living
things are continually recycled by way of the CO_2 in the air. The carbon
cycle has many subtleties and sophistications when considered in detail
(methane and other volatile organic compounds also play a significant
part) but in crude outline the description just given is true: living things,
on this planet, depend on the cycling of carbon between living matter
and atmospheric CO_2, a process powered by the solar energy which has
been trapped by photosynthetic organisms. The other major biological
elements, nitrogen, oxygen and sulphur, are subject to comparable cyc-
lic processes, and the most important, from both ecological and economic viewpoints, is the nitrogen cycle.

   The nitrogen cycle can be drawn in several ways, of which that in Fig.
1.1 is a simple example. Essentially it symbolizes the transformations
undergone by the element nitrogen (N) on this planet through the
agency of living things. The branch and lower sector show the synthesis
of nitrogenous living matter (principally protein) from inorganic nitrogen compounds (nitrate, nitrite and ammonium ions) during growth of
plants and their consumption by animals, followed by their return to
the soil as a result of decay and putrefaction of plant and animal material. The upper sector shows the loss of nitrogen to the atmosphere
from nitrates, and its return to the cycle by the process known as
nitrogen fixation. The steps of the cycle will be discussed individu-
ally later in this chapter; an important general point is that, in

1

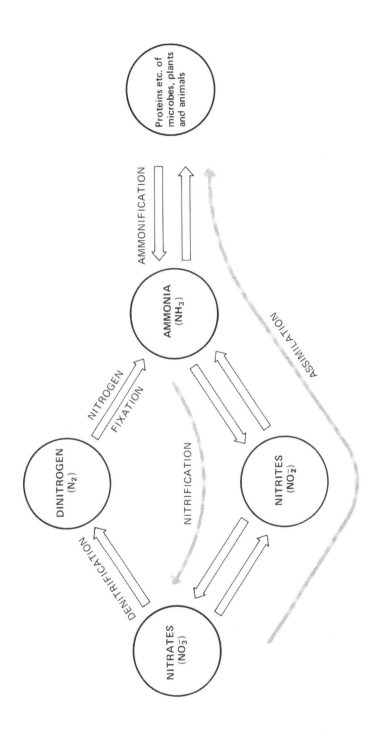

Fig. 1.1. The biological nitrogen cycle.

nearly all agricultural areas of this planet (areas where shortages of sunlight and/or water supply do not limit crop growth), biological productivity is determined by the availability of inorganic nitrogen in the soil. This means that in all but highly sophisticated agricultural communities, the rate at which the cycle turns determines biological productivity.

The element nitrogen is an essential constituent of all living things – the proteins and nucleic acids are their major nitrogenous constituents but most other biological materials contain some nitrogen atoms – and one can calculate that the plants and animals of the soils and waters of this planet together contain roughly 1.5×10^{10} tonnes of N. Each year the nitrogen cycle transforms roughly one-fifteenth of this (about 10^9 tonnes N; this number is difficult to calculate and may be low by a factor of 2 or more). Much of the cycling is by way of the lower part of the cycle, but between 2 and 3×10^8 tonnes N per year (a rather more reliable number by virtue of the acetylene test described in section 2.2) passes through the upper cycle. In practice, this means that input of inorganic N to the biosphere by the process called nitrogen fixation is rate-limiting for biological productivity on most areas of the sea or land surface of this planet. There are two major exceptions to this statement. Firstly, in undisturbed localities such as virgin forest or savannah biological fixation provides an adequate N-input and other factors (water, pH, phosphate, etc.) become limiting. Secondly, in farmlands where N, supplied by putrefaction processes or application of nitrogenous fertilizer, exceeds the N demand of the crops, other nutrients (K, P, S or, rarely, a micronutrient such as Mg or Mo) become limiting.

1.2 Nomenclature

The term 'nitrogen' really refers to the N atom, and the gas which composes 80% of our atmosphere, N_2, is correctly termed dinitrogen. This term will be adopted here but, since the more colloquial 'nitrogen' is still widely used for the process of nitrogen fixation, I shall continue to use it in that particular context. 'Fixed nitrogen' will refer to any compound of N that is not N_2 but may have originated therefrom. Other technical terms will be explained as they arise; those which specially concern the nitrogen cycle are discussed in the next five sections.

1.3   Assimilation of nitrogen

This is a general term for the biological conversion to organic nitrogen of fixed nitrogen compounds, illustrated as nitrate, nitrite or ammonia in Fig. 1.1. A growing plant or microbe in soil assimilates nitrogen as nitrate, converting it into protein, nucleic acids and minor nitrogenous components of the cell; an animal assimilates amino acids, building these into protein and other biological polymers. The pathways of nitrogen assimilation in living things together constitute a complex subject which could easily occupy a book of this size on its own. For present purposes, discussion will be restricted to the compounds illustrated in the cycle.

All plants and many bacteria reduce nitrates to ammonia by way of nitrites; the ammonia is then incorporated into nitrogenous biopolymers. This process is called *assimilatory* nitrate reduction, because the nitrogen of nitrate is assimilated into protein. It is the major process whereby nitrogen is incorporated into plant, and hence animal, material. The ammonia formed is assimilated primarily by incorporation into amino acids, and the commonest reaction of this kind is formation of glutamic acid from 2-oxoglutaric acid, a normal intermediate in carbon metabolism. The enzyme responsible for the reaction is called glutamate dehydrogenase and the reaction [1.1] is reversible:

$$NADH + HOOC.CH_2.CH_2.CO.COOH + NH_3 \rightleftharpoons HOOC.CH_2.CHNH_2.COOH + H_2O + NAD \qquad [1.1]$$

$$\text{2-oxoglutarate} \qquad \text{ammonia} \qquad \text{glutamate} \qquad \text{water}$$

(NADH and NAD are the reduced and oxidized form of a biological reducing agent).

Glutamate is a key substance in the formation of other amino acids and of protein. An important point is that the enzyme is not only reversible but has a rather low affinity for ammonia: glutamate dehydrogenase is relatively inefficient at picking up the ammonia molecule. Bacteria are frequently obliged to exist in conditions of shortage of ammonia, as well as of precursors such as nitrite or nitrate. In such circumstances, many can use a different series of reactions to assimilate the little ammonia available, one which makes use of two enzymes. Essentially, the first enzyme, glutamine synthetase, binds ammonia by reaction [1.2] to an existing molecule of glutamate to form glutamine:

$$\text{HOOC.CH}_2\text{.CH}_2\text{.CHNH}_2\text{.COOH} + \text{NH}_3 + \text{ATP} \rightarrow \text{NH}_2\text{.CO.CH}_2\text{.CH}_2\text{.CHNH}_2\text{.COOH} + \text{H}_2\text{O} + \text{ADP} + \text{P} \quad [1.2]$$

glutamate   glutamine

(ATP and ADP are the charged and discharged forms of molecules responsible for the transfer of biological energy; inorganic phosphate (P) is released in the conversion of ATP to ADP, they are discussed further in section 2.2).

Glutamine synthetase is very efficient at picking up ammonia, but has the disadvantage as far as the economy of the microbe is concerned that one molecule of ATP, the biological 'quantum' of energy, is lost for each molecule of glutamine formed. Put in its simplest terms, the organism has to use up energy, as ATP, to make glutamine synthetase work. The glutamine so formed then reacts with 2-oxoglutarate, with the aid of a second enzyme, glutamate synthase, to form two glutamate molecules as in reaction [1.3]:

$$\text{NADH} + \text{NH}_2\text{.CO.CH}_2\text{.CH}_2\text{.CHNH}_2\text{.COOH} + \text{HOOC.CH}_2\text{.CO}_2\text{CO.COOH} \rightarrow 2\text{HOOC.CH}_2\text{.CH}_2\text{.CHNH}_2\text{.COOH} + \text{NAD}$$

glutamine   2-oxoglutarate   glutatamate   [1.3]

By reactions [1.2] and [1.3], then, the organism converts one glutamate molecule into two, assimilating ammonia efficiently but expending ATP. Thus microbes have two pathways of ammonia assimilation: one 'cheap' in terms of ATP but inefficient, for use when ammonia (or a precursor such as nitrate) is plentiful; another, 'expensive' but highly efficient, for use when supplies of ammonia are short. Once glutamate has been formed, the inorganic nitrogen of nitrate or ammonia has entered an organic compound on its way to being built into a living organism. Other routes exist whereby N can enter organic combination, but glutamate formation is the principal reaction.

1.4   Ammonification

Inorganic nitrogen is returned to the cycle from organic matter as a result of autolysis, decay and putrefaction of biological material, and the principal form in which it appears is as ammonia. The general process is referred to as ammonification. The enzyme glutamate dehydrogenase, mentioned in the previous section, when working in reverse, is typical of the enzymes responsible for this process; as in reaction [1.4] it forms a keto acid from an amino acid:

$$\text{NAD} + \text{HOOC.CH}_2.\text{CH}_2.\text{CHNH}_2.\text{COOH} + \text{H}_2\text{O} \rightleftharpoons \text{HOOC.CH}_2.\text{CH}_2.\text{CO.COOH} + \text{NH}_2 + \text{NADH} \quad [1.4]$$
$$\qquad\qquad\;\; \text{glutamate} \qquad\qquad\qquad\qquad\qquad \text{2-oxoglutarate}$$

A moment's inspection will show that reaction [1.4] is the reverse of reaction [1.1]. Ammonia can also be released from amino acids by a hydrolytic reaction (the enzymes responsible are called deaminases) and from amides by deamidases; a familiar one is urease, which hydrolyses the urea of urine, causing the familiar smell of animal houses and imperfectly cared-for toilets. Details are inappropriate here, the essential point being that breakdown of proteins and other nitrogenous organic matter almost always leads to the formation of free ammonia.

1.5   Nitrification

The ammonification step of the cycle contributes free ammonia to the biosphere. Two classes of bacteria exist which live by oxidizing the ammonia to nitrate. Both are unusual forms of life: just as plants use sunlight to convert CO_2 to organic matter, these bacteria use the energy of ammonia oxidation to 'fix' CO_2. Like plants, the majority cannot use external sources of organic matter (such as glucose): for reasons which microbiologists still do not wholly understand, they are obliged to make their own organic matter from CO_2. Organisms of this kind are called *autotrophs*. Plants belong to the class of autotrophs called *phototrophs*, because they use light energy to fix CO_2; these bacteria are called *chemotrophs*, because they couple a chemical reaction to CO_2 fixation.

The biological oxidation of ammonia to nitrites and nitrates is called nitrification and the microbes responsible are called nitrifying bacteria. They can be divided into two main groups. *Nitrosomonas* is typical of the principal six or so genera which oxidize ammonia to nitrite (see Fig. 1.1); *Nitrobacter* represents the rather fewer genera which oxidize nitrite to nitrate. Both classes of bacteria are very widespread but difficult to isolate and culture in the laboratory. From an ecological point of view they have two important functions, one beneficial and one deleterious. Ammonia, though an adequate plant nutrient, is less effective than nitrate and most plants prefer nitrate. The nitrifying bacteria perform a valuable function in rendering the nitrogen of ammonia more readily available to plants. On the debit side, ammonia is well retained by soil, whereas nitrates are readily washed away, so nitrification can result in loss of nitrogen from that zone of soil which is readily accessible to the roots of plants: the rain-washed surface layers. An example in which this process is economically important is the leaching of artificial fertilizer. Much of the artificial fertilizer used in agriculture is free ammonia or an ammonium salt; nitrification can lead to loss of nitrogen from the fertilizer farmland to neighbouring soils and waters, occasionally causing contamination of drinking water. Ammonia from natural ammonification processes (e.g. a compost heap rich in nitrogen) can also become converted to nitrate and cause similar problems; agricultural chemicals are available which restrict this process by inhibiting growth of nitrifying bacteria.

1.6   Denitrification

The process of assimilatory nitrate reduction was mentioned in section 1.3. Some microbes can use nitrates as a substitute for oxygen in respiration. The process is called *dissimilatory* nitrate reduction, because it is associated with the dissimilation (oxidation) of organic matter. Often the nitrate is converted to nitrite (advantage is taken of this reaction in the curing of bacon: the nitrite formed from nitrate reacts with meat protein, turning it red, altering its taste and protecting it from decay). Much of the nitrite formed then enters the assimilatory pathway of the nitrogen cycle (Fig. 1.1). There exist, however, several groups of bacteria which reduce nitrate to dinitrogen, and these bacteria are common in such environments as sewage treatment plants, compost heaps and rich soil. Examples are found in the genera *Pseudomonas*, *Micrococcus* and *Thiobacillus*. The dinitrogen is released and becomes part of the atmosphere. An intermediate in the biological reduction of nitrate is nitrous oxide, N_2O; escape of this gas, followed by decomposition to N_2, can also return N_2 to the atmosphere. These processes cause a net loss of biological nitrogen from the world's soils and water; reliable data for their global scale are not available, but a figure in the region of 2×10^8 tonnes N per year lost as N_2 is probably reasonable.

1.7   Nitrogen fixation

The loss of nitrogen to the atmosphere recorded above is substantial and, even though the number may be out by a factor of 2, it would soon bring life on this planet to a stop as a result of N-starvation. In fact, this loss is compensated for by the process called nitrogen fixation and, as pointed out in section 1.1, the nitrogen economy of this planet is such that, in most fertile areas, the rate of N_2 fixation determines biological productivity.

The ability to fix N_2 is restricted to the most primitive living things, the bacteria and even within this group the property is by no means universal. Bacteria capable of fixing N_2 are called diazotrophs. The first product of fixation is ammonia, as the cycle in Fig. 1.1 illustrates, but it is important to emphasize that the ammonia is nearly always assimilated as rapidly as it is formed. From an ecological (or agricultural) viewpoint, the most important diazotrophs are those which fix in association with a plant (see Chapter 5), because they supply fixed N just where it is needed: close to the plant's roots. As stated in section 1.1, an annual input of some 2×10^8 tonnes of N takes place on this planet each year, of which biological processes contribute the bulk.

Nitrogenous fertilizer is today mostly made from atmospheric dinitrogen by a reaction [1.5] which is known as the Haber process, essentially a catalytic reduction of dinitrogen to ammonia:

$$N_2 + 3H_2 \rightarrow 2NH_3 \qquad [1.5]$$

The hydrogen is generated from natural gas and the reaction requires high pressures and a moderately high temperature to be efficient. According to Dr R. W. F. Hardy, a leading authority on these matters, the world's industrial output of nitrogenous fertilizer increased an astonishing 27-fold between 1950 and 1990, when it reached some 8×10^7 tonnes N/year: approaching 50% of the total global input of N into the terrestrial biosphere. Lightning and ultraviolet radiation cause the formation of nitrogen oxides, particularly in the upper atmosphere, and combustion (particularly the internal combustion engine) and electrical sparking generate oxides of nitrogen at ground level (in cities, the nitrogen oxides from car exhausts can be a serious atmospheric pollutant). These gases become washed into soil by rain and dew, contributing to the total

Table 1.1. *Estimated N-input into plant crops in 1971–2*

Source of N	Millions of tonnes of N			
	UK	USA	Australia	India
Fixation in legumes	0.4	8.6	12.8	0.9
Fixation by free-living microbes	<0.05	1.4	1.0	0.7
Taken up from N fertilizer	0.6	4.9	0.1	1.2

N-input from atmospheric dinitrogen. Figures for this non-microbial input are not easily obtained, but authorities agree that such processes are unlikely to contribute more than 10%. It is reasonable, therefore, to accept that some 40% of the input of the world's soil and water nitrogen is today supplied by nitrogen-fixing bacteria in one form or another.

It is important to put this matter into perspective. The amount of N present on this planet as dinitrogen is vast: Something like 4×10^{15} tonnes in the atmosphere and a surprising 2×10^{16} tonnes bound in sedimentary and primary rocks beneath the surface. None is accessible to plants until it is fixed, principally by microbes; in many circumstances the microbes have to die and decompose – ammonification has to take place – before the plants can gain access to the fixed nitrogen. In symbiotic systems, in which nitrogen-fixing bacteria are intimately associated with plants, the transfer of fixed nitrogen from microbe to host is much more rapid and efficient, so symbioses of the kind discussed in Chapter 5 are much the most important agents of biological fixation as far as global biological (and agricultural) productivity are concerned. Nonetheless, free-living microbes make a serious contribution, particularly in poor or unfertilized soils. Many cyanobacteria fix nitrogen and can be the primary source of N-input in the tundra, in the sea, on stony or devastated land. This matter is discussed further in Chapter 4. In highly developed agricultural countries, where ammonia fertilizer is available and cheap, one might expect the biological processes to be unimportant. Yet agronomists have evidence, indicated in Table 1.1, that only about one-third of the nitrogen produced, in plants, by agriculture in the United States in 1971–2 entered the soil as artificial fertilizer; the rest appeared by biological fixation. In Australia, where the exploitation of nitrogen-fixing plants in agriculture is highly developed, chemical fertilizer accounted for less than 1% of crop nitrogen. It is curious that, in India, where soil is often poor, crop yields low, yet demand intense, about 40% of the crop nitrogen nevertheless originated from industrial fertilizer. The possibilities of exploiting the biological process to alleviate low levels of nutrition in developing countries are obvious.

The principal nitrogen-fixing systems useful in world agriculture are the legumes such as peas, beans, soybeans, chickpeas, lupins and lucerne (see Table 5.1). They all involve association of a plant with a group of related bacteria loosely termed 'rhizobia' (the best-known belong to the genus *Rhizobium*). Characteristically the bacteria colonize zones of the roots in little excrescences called nodules (see Chapter 5 and Fig. 5.1). Such plants can be made use of in three principal ways. Direct use as food is most familiar: peas and beans are a well known and regular part of most people's diet; rather less well known is the now widespread use of protein from soybeans in the food industry. Conversion to meat products via cattle fodder is the second use: pulses and lucerne (alfalfa) are increasingly used as cattle food supplements, both as the plant and after conversion to silage. The third major use is as green manure: clovers, melilot and lucerne are grown and ploughed in to upgrade poor or spent agricultural land, particularly in communities which practise intelligent crop rotation. Use of clover to upgrade the fertility of Australian soil led to dramatic improvements in the 1940s. Sometimes a given plant system can serve two purposes: certain non-toxic varieties of lupins have been recommended for agricultural use in Western Australia because their seeds prove to be a useful high-protein grain for animal feed and the plants can be ploughed in as a green manure after harvest.

1.8   Nitrogen fixation and energy resources

An important feature, particularly to developing countries, of biological nitrogen fixation is that the energy for the conversion of dinitrogen to ammonia generally comes from sunlight. Legumes, for example, supply the products of photosynthesis to the bacteria in their nodules, who use them for fixation. Cyanobacteria use sunlight both to grow and to fix nitrogen. The Haber process requires considerable energy to prepare hydrogen from natural gas and rather less energy to compress the re-

agents and heat them. In addition, quite a complex industrial installa-
tion is required, so production must be localized, and a further consumption of energy (as fuel for lorries, trains, tractors and for packaging)
is required to transport the ammonia from the factory to where it is
needed. Transport is far from a trivial question: as long ago as 1964
scientists calculated that transport facilities equivalent to the world's
shipping tonnage at that time would be needed by the year 2000, solely
to transport fixed nitrogen from factories to farms, if the rate of population growth at that time was sustained. The rate of population growth
has decreased slightly, in global terms, but it is now clear that the world's
energy supplies must be conserved. Petroleum products, particularly,
have escalated in cost and will rise again. Petroleum costs do not
seriously affect production costs of ammonia fertilizer, because the
primary energy source is methane, but they affect secondary costs such
as packaging and transport, and costs of natural gas tend to rise with
other energy costs. A dramatic rise in the price of ammonia fertilizer took
place in the early 1970s (4.5-fold between 1972 and 1975) but, though
energy costs played a small part, the major cause was a steadily increasing world demand for fertilizer which outstripped supply. More factories
can be built, and more ammonia fertilizer can be made, but in the long
run the energy costs for methane, transport and so on are likely to
become prohibitive. The global need to limit use of the fossil fuels coal,
methane and petroleum, provides an additional reason for exploiting
biological nitrogen fixation, over and above the escalating demand for
food.

2   The enzyme

2.1   Nitrogenase

Diazotrophic bacteria conduct a reaction which a chemist would consider 'difficult'. The dinitrogen molecule is normally very unreactive: red hot magnesium, or a catalyst at elevated pressures and temperatures as in the Haber process, are normally necessary to make N_2 reactive. This chemical inertia is mainly due to the very stable triple bond which links the two atoms in the structure $N{\equiv}N$. Diazotrophs make the N_2 molecule reactive at ordinary temperatures and pressures and, in addition, they do so in water under an atmosphere of oxygen – both substances which would interfere drastically with such reagents as red hot magnesium or the Haber system! The biological catalyst, or enzyme, responsible for this process must have some exceptional chemical properties, and some of these are still being revealed.

In fact a large number of nitrogen-fixing enzymes exist, distributed among over 100 species of nitrogen-fixing bacteria, but they are biochemically so similar that it is useful to refer to the whole family of enzymes as 'nitrogenase' even though that name does not conform to systematic enzyme nomenclature. In 1960 a group of workers in the Central Research laboratories of Dupont de Nemours (a large chemical firm in the USA) obtained the first solution of nitrogenase from *Clostridium pasteurianum*. This is an anaerobic bacterium common in soil and compost heaps; it is unable to grow in air yet was one of the first nitrogen fixers to be discovered. The extract which the Dupont workers made was a sort of juice prepared by resuspending dried bacteria in aqueous salt solutions and centrifuging; it converted N_2 to ammonia (NH_3), a reaction proved by using the isotope of nitrogen $^{15}N_2$ and isolating $^{15}NH_3$. A critical point, which later proved to have an importance far beyond the isolation of the enzyme, is that the researchers had to conduct all their manipulations in the absence of air: nitrogen-fixing activity was completely and rapidly destroyed if oxygen reached the extract.

Nitrogenases have now been extracted from some 30 diazotrophic microbes, as biochemists developed ways of handling enzymes in the absence of air, excluding air from electrophoresis systems, chromatography columns, centrifuges and so on. Nitrogenases differ slightly from species to species and there seem to be three main classes (see section 2.4). Purified nitrogenases have been made from about eight microbes, including legume nodules.

Table 2.1 gives some properties of the enzyme purified from three different species of diazotroph. *C. pasteurianum* is the anaerobe from which the first extract was made; *Azotobacter chroococcum* is a very different microbe, which can grow only in air and which is widespread in soils, particularly chalky or sandy ones; *Klebsiella pneumoniae* is also a common soil bacterium, physiologically intermediate but not related to either of the other two: it can grow either with or without air but it can fix nitrogen only in the absence of air.

The first thing that Table 2.1 shows is that nitrogenase is not one protein. On purification it separates into two absolutely distinct proteins, both dark brown in colour, neither active without the other. The big one consists of four subunits (polypeptide chains) stuck together in some way, two of the subunits being different from the other two. The smaller protein consists of only two subunits, apparently identical. Both contain iron atoms, the amount per molecule of the large protein being substantial: the smaller one having four per molecule. In both cases the iron is accompanied by essentially the same number of sulphur atoms. (These sulphur atoms are termed labile sulphur in the table; this means sulphur which can be displaced as H_2S by weak mineral acid. It is known to be sulphur actually attached to the iron atoms.) The big protein also has two atoms of the metal molybdenum per molecule. (Iron and molybdenum belong to the class of elements which chemists call the 'transition metals', a rather ill-defined class including group VIII of the periodic table and some in groups VII and VI.)

Table 2.1 also shows that the two proteins are much alike no matter which microbe they come from. Their similarity is so great that, in the laboratory, one can take the big protein from *Klebsiella pneumoniae*

Table 2.1. *Some properties of the two component proteins of nitrogenase purified from three species of bacteria*

Protein	Cp1	Kp1	Ac1	Cp2	Kp2	Ac2
Molecular weight (mw)	220 000	218 000	227 000	55 000	66 700	64 000
Number and mw of sub-units	$\begin{cases} 2\times 50\,700 \\ + \\ 2\times 59\,500 \end{cases}$	$2\times 51\,300$ + $2\times 59\,600$	Probably 2×2 types 2	$2\times 27\,500$	$2\times 34\,000$	$2\times 31\,000$
Mo atoms/molecule[a]	2	2	2	0	0	0
Fe atoms/molecule[a]	22–24	30–35	21–25	4	4	4
Labile sulphur/molecule[a]	22–24	>18	18–22	4	4	4
Half-life in air (minutes)	Short	10	10	Very short	0.75	0.5

Cp, *Clostridium pasteurianum*; Kp, *Klebsiella pneumoniae*; Ac, *Azotobacter chroococcum*; 1 and 2 refer to the larger and smaller proteins irrespective of origin.
[a]Data rounded to nearest whole numbers.

(Kp1) and mix it with the small protein from *Azotobacter chroococcum* (Ac2) and get a fully active 'hybrid' nitrogenase enzyme. The converse mixture (Kp2 + Ac1) is also fully active, but crosses with the clostridial proteins to give only weak, if any, activity.

Most impressive is the readiness with which the proteins are destroyed by oxygen, particularly the smaller protein. This is true even when the proteins are purified from *A. chroococcum* which, because of its aerobic habit, might have been expected to have more oxygen-tolerant nitrogenase. In fact, there is a sense in which it has, because crude extracts of *A. chroococcum* (made by squashing the bacteria in an appropriate apparatus and removing the cell envelopes) contain the nitrogenase in small, subcellular aggregates in which the proteins are quite tolerant of oxygen. Only when the aggregate is broken up and the proteins purified from it do they show the high oxygen sensitivity illustrated in Table 2.1.

2.2   Substrates for nitrogenase

Nitrogenase, then, consists of two proteins. It reduces dinitrogen to ammonia: one molecule of N_2 is converted quantitatively to two molecules of the hydride NH_3. Nitrogen can form two other hydrides: the unstable di-imide ($HN{=}NH$) or the more stable hydrazine ($H_2N{-}NH_2$), but careful tests for these compounds using the isotope $^{15}N_2$ have given no sign of them as free intermediates. Yet good evidence exists for an enzyme-bound hydride of N_2 being formed during the reaction. The conversion of N_2 to NH_3 is a reduction (like the Haber process) and a reducing agent corresponding to the hydrogen of the Haber process is needed, a matter which will be discussed in the next chapter. In the earliest experiments it was present in the cell extracts. Fortunately for research, the laboratory chemical sodium dithionite (earlier known as sodium hydrosulphite), $Na_2S_2O_4$, works and it is now widely used as a reductant in enzyme studies.

In order to function at all nitrogenase needs adenosine triphosphate (ATP), the biological 'quantum' of energy alluded to in section 1.3. This substance is present in all living tissue, and is the principal molecule whereby chemical energy is mobilized for biological processes. ATP is hydrolysed to adenosine diphosphate (ADP) with a net loss of chemical energy which can be used by living cells for all sorts of biological processes; when nitrogenase functions it converts ATP to ADP, the energy released being somehow used in the fixation process. In the test tube, 16 ATP molecules are consumed to convert one N_2 to two NH_3 molecules. In physiological terms, nitrogen fixation is an expensive process, because the organism must use food (glucose for example) to make that ATP; it usually regenerates ATP from ADP by processes which are well understood but which need not concern us here. Historically, the involvement of ATP took rather a long time to discover, because ADP is an inhibitor: if ADP is allowed to accumulate, it blocks some step in the reaction and everything comes to a standstill. Research workers today usually use an artificial ATP-regenerating system to overcome this problem.

Another essential ingredient for nitrogenase is the magnesium ion, Mg^{2+}. If this is completely absent, the enzyme will not function. Quite what the Mg^{2+} is doing is still not understood, but one fact is already clear: ATP does not react with nitrogenase on its own; it has first to become the mono-magnesium salt (often written ATPMg, though this is in no sense a correct chemical formula) to perform its function.

The complications of nitrogenase are not over yet. If dinitrogen is present, it is reduced to ammonia. But if it is not present (with an active preparation under argon, for example), the enzyme then reacts with water, forming gaseous hydrogen. What happens is that hydrogen ions (H^+) of water become reduced and join together to form H_2 molecules: the enzyme is a powerful enough reducing agent to decompose water. In fact, this reaction actually accompanies nitrogen fixation when the enzyme is used in the test tube and there is evidence that it also takes place in living bacteria, but that some have ways of recycling the hydrogen formed. The evolution of hydrogen requires a reducing agent, ATP and Mg^{2+}, of course, just like the nitrogen fixation reaction proper.

Nitrogenase reduces the triply bonded $N\equiv N$ molecule to ammonia. A number of molecules exist which are a little like dinitrogen in that they are small and have a triple bond. Examples are:

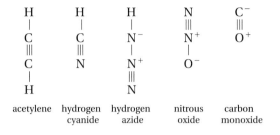

acetylene	hydrogen cyanide	hydrogen azide	nitrous oxide	carbon monoxide

The first four substances are indeed reduced by nitrogenase, provided ATP and Mg^{2+} are available, but the fifth, carbon monoxide, is not reduced. However, carbon monoxide certainly interacts with the enzyme since it blocks the reduction of all the others, and of N_2, if it is present. Therefore, it is thought to bind to whatever is the site on the enzyme which takes up N_2, but to be incapable of being reduced. It is interesting, however, that carbon monoxide does not interfere with the hydrogen evolution reaction mentioned in the previous paragraph.

Acetylene is a particularly important substrate because the product, ethylene, can be detected rapidly and with great sensitivity by gas chromatography. This reaction is the basis of the now famous acetylene test for nitrogen fixation, because it takes place not only with enzyme preparations but also with living bacteria, plants or excised nodules, soil or water samples. One has only to expose a nitrogen-fixing system to acetylene, in appropriate conditions of temperature, oxygen content and so on, and, sometimes within minutes, the rate of ethylene formation will provide a measure, usually quantitative, of the nitrogen-fixing capacity of the system under test. This test is quicker, cheaper and much easier than the more reliable tests using $^{15}N_2$. Its application has not only revolutionized the study of nitrogen fixation in nature, but permitted the rapid progress in understanding the biochemistry and genetics of nitrogen fixation which took place in the 1970s and 1980s. However, the acetylene test must be used with careful controls. In particular, it can fail with bacteria whose metabolism involves methane (see section 4.3) or carbon monoxide (see section 2.6); and damaged or ripening plant tissue can mislead by forming ethylene independently of acetylene. Wise researchers always confirm innovative discoveries with $^{15}N_2$.

Studies of the reduction of bizarre substrates such as methyl isocyanide or methyl acetylene give circumstantial evidence that metals are involved in N_2 reduction: the reduction products formed by the enzyme prove to be those chemists would expect if the substrate first became bound to a transition metal atom such as iron or molybdenum. Further evidence in this direction is provided by spectroscopic and mechanistic studies recounted in the next section, but it will be helpful here to summarize the several special properties of nitrogenase:

1 consists of two proteins
2 is destroyed by oxygen
3 contains the transition metal atoms iron and molybdenum
4 needs Mg^{2+} ions to be active
5 converts ATP to ADP when functioning
6 is inhibited by ADP
7 reduces N_2 and several other small, triply bonded molecules
8 reduces hydrogen ions to gaseous hydrogen, even when N_2 is present.

A species of diazotrophic bacteria exists whose nitrogenase is apparently aberrant in properties 2, 6 and 7 of this list – see section 2.6. However, the properties listed above apply to all other nitrogenases so far studied and make it clear that the major question of how nitrogenase works needs answers to a number of subsidiary questions. For example, how and in what proportion do the two proteins interact? Why and where does ATP react? Why are the metal atoms there? Where do dinitrogen or other substrates bind? Is hydrogen evolved from the same or a different site? Present knowledge cannot completely answer any of these questions, but the study of nitrogenase has advanced sufficiently to allow at least a provisional idea of the way it functions.

2.3   The two proteins and the nitrogenase complex

Proteins which contain iron, such as the haemoglobin of blood, have visible and ultraviolet spectra which give invaluable information about their behaviour. Comparable spectra of nitrogenase proteins are, however, rather uninformative. There exists, however, a form of spectroscopy (Mössbauer spectroscopy) which uses γ-radiation instead of light and which works very well with chemicals containing iron. Such spectra of nitrogenase proteins have been particularly revealing as far as the molybdoprotein is concerned. They showed that this protein could exist in three different states, distinguished by the proportion of the iron atoms which were in the reduced (ferrous) condition in the molecule. The most oxidized form is probably physiologically unimportant; the intermediate form is that in which the protein is usually isolated; the most reduced form appears only when the other protein is present together with ATP, Mg^{2+} and sodium dithionite. In other words, the most reduced form appears only when all the components necessary for functioning nitrogenase are supplied, and is, in fact, the predominant form in the functioning enzyme.

Proteins with iron in them have unusual magnetic properties. A very sensitive kind of spectroscopy which exploits these properties is called electron paramagnetic resonance, usually abbreviated to epr. The intermediate form of the molybdoprotein shows a very clear and distinct epr absorption but the other two forms do not. If the molybdoprotein is treated with acid, a fragment can be obtained from it called 'FeMoco': a small, unstable molecule which contains iron, sulphur and molybdenum and which still has the epr signal. Its chemical structure is discussed in section 2.4. There exist mutant diazotrophs (e.g. *nifB* mutants of *Klebsiella*, see Chapter 6) which cannot fix N_2 because their molybdoprotein is defective: it lacks FeMoco. It is actually possible to reconstitute active molybdoprotein from a FeMoco solution and *nifB* nitrogenase. Thus the FeMoco fragment is important to nitrogenase function. But genetics has revealed something else. Another mutant, called *nifV* (see Chapter 6) makes nitrogenase which can reduce acetylene but not N_2 and which has altered reactions with other substrates. If FeMoco from *nifV* nitrogenase is added to the inactive *nifB* protein, one reconstitutes the *nifV* type of nitrogenase. These roundabout but very useful experiments, combining genetics and biochemistry, tell us that FeMoco is, includes, or is part of, the N_2-binding site of nitrogenase. Thus the molybdoprotein is the true 'dinitrogenase' and the iron protein has some other function. Scientists are tempted to the view that the N_2-binding site is a molybdenum atom, but there is no really firm evidence for this view.

If the molybdoprotein binds N_2, what does the smaller iron protein do? Epr has been very useful here, too, because this protein can exist in oxidized or reduced forms and the latter has an epr spectrum. If ATPMg is added, the epr spectrum of the reduced form undergoes a substantial change, a change accompanied by an increase in chemical reactivity: the protein is even more easily destroyed by oxygen, loses its iron more easily, reacts with other reagents more rapidly. Effectively it becomes a more powerful reducing agent, one capable of generating the most reduced form of the molybdoprotein. Kinetic experiments show that two ATP molecules are hydrolysed to ADP each time an electron is transferred to the molybdoprotein.

Reduction of N_2 to $2NH_3$ requires six electrons, equivalent to 12 ATP molecules. Yet the enzyme consumes about 16 ATPs. The reason for this became clear in the early 1980s: reduction of N_2 is always accompanied by the formation of one molecule of H_2, which means the consumption of a further two electrons or four ATPs. Why is this by-product formed at all? Kinetic and chemical studies are converging to suggest that the FeMoco site, before it can bind N_2, must bind two H^+ ions as hydride groups, and that these can then be displaced by N_2. The enzyme reaction should be written formally so:

$$N_2 + 8H^+ + 16ATP \xrightarrow{8e} 2NH_3 + H_2 + 16ADP \qquad [2.1]$$

The direct involvement of hydrogen in the functioning of nitrogenase ties in with an observation that had been made in the 1930s, during some of the earliest studies on the biochemistry of nitrogen fixation. Hydrogen exerts an inhibitory effect on nitrogen fixation by root nodules and by *Azotobacter*. Also, in appropriate conditions, nitrogen-fixing systems catalyse a reaction in which a mixture of hydrogen (H_2) and its heavy isotope deuterium (D_2) interact to give the hybrid molecule hydrogen deuteride (HD). One must add, however, that the precise way in which nitrogenase promotes these interactions is still not clear.

It is today possible to measure very fast chemical reactions, and such studies, investigating the stages by which nitrogenase gets started, show that H_2 evolution does indeed precede binding of N_2 and that a bound intermediate, something like a N_2H_4 complex, occurs *en route* to NH_3. Most fascinating is the discovery that the enzyme is remarkably slow: it takes 1.25 sec for a molecule of enzyme to form two of NH_3. This slowness arises in part because the two proteins have to come together and separate eight times to reduce one N_2 molecule. It is their separation which is the slowest step.

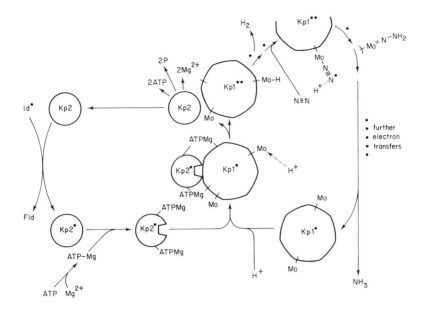

Fig. 2.1.   A scheme for the action of *Klebsiella* nitrogenase. Kp1 is the larger protein, containing iron and molybdenum atoms; Kp2 is the smaller protein with iron atoms only. Fld is flavodoxin, which donates an electron (a black dot) to iron atoms in Kp2. ATP reacts with magnesium ions to produce a compound which activates the reduced form of Kp2. Meanwhile Kp1, with a 'spare' electron among its iron atoms, has bound a hydrogen ion from water at a transition metal atom depicted as molybdenum (see text). Activated Kp2 joins Kp1 carrying substrate to form a complex, assisted by ATPMg, within which an electron from the iron atoms in Kp2 is transferred to the iron atoms in Kp1, prior to reaching the bound substrate. After two such electron transfer events, the two bound hydrogen atoms are displaced by N_2 and released as hydrogen gas. Further electron transfers lead to hydrogen ions binding to the bound N_2 and so by way of at least one metal-bound intermediate stage (depicted as —Mo=N—NH_2) the product, NH_3, of enzyme action is released. Each electron transfer requires a new electron from Kp2 and each time the transfer takes place two molecules of ADP are formed from the ATP. The scheme is one of several that could be proposed, based on the experimental findings that Kp2 is concerned with ATP consumption and electron transfer, Kp1 with substrate binding and reduction.

In summary, then, present evidence is that, when nitrogenase works, the molybdoprotein has the dinitrogen molecule bound, after displacement of a hydrogen molecule, at some metal atom which may well be molybdenum, and the smaller protein has ATP bound as its mono-magnesium salt. The fixation process involves reduction of iron in the molybdoprotein at the expense of iron in the smaller protein; as a consequence ATP becomes converted to ADP and an electron ends up on the N_2. This is neutralized by a hydrogen ion, forming an N_2—H bond. A sequence of such steps ultimately yields NH_3. These reactions are brought together in a scheme in Fig. 2.1; this must be taken as suggestive rather than as representing established fact but it indicates a substantial advance in understanding nitrogenase since the first extracts were obtained in 1960.

2.4   FeMoco and the structure of nitrogenase

In 1992 the three-dimensional chemical structure of Av1, the molybdo-protein of the nitrogenase purified from *Azotobacter vinelandii* (Fig. 2.2), was established by X-ray diffraction. It confirmed less direct evidence that the protein consisted of two pairs of distinct subunits, with the FeMoco group buried in a cleft in one of them, relatively inaccessible to water molecules. FeMoco itself, which had defied conventional structural analyses, was revealed to be a complex inorganic cluster containing seven iron and one molybdenum atoms linked by sulphur atoms, itself bound into the protein by way of a cysteine and a histidine group. The other iron–sulphur groups within the protein were present as pairs of clusters composed of four iron and four sulphur atoms each and their locations were clarified; homocitrate, a group which was known from studies with the *nifV* mutant to be closely involved with FeMoco, was revealed to be bound to the molybdenum atom. Figure 2.3 is a sketch of the FeMoco group as implied by the X-ray analysis. The structure of Av1 resolved several questions about the way in which nitrogenase works, including the theoretical need for the active site to limit access of hydrogen ions, and was wholly consistent with the sort of mechanism proposed in Fig. 2.1, with the added detail that the non-FeMoco iron–sulphur clusters were well placed to receive electrons from Av2 and pass them by intramolecular transfer to FeMoco.

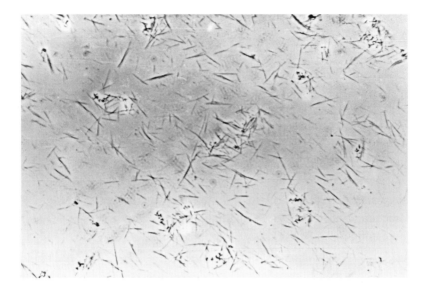

Fig. 2.2.   Crystals of the molybdoprotein of *Azotobacter vinelandii*. The needle-like crystals are 10–20 μm long.

2.5   The alternative nitrogenases

So far only the conventional Mo-containing nitrogenases of diazotrophs have been discussed. In the early 1980s, however, Professor Paul Bishop and his colleagues in the USA obtained evidence that *Azotobacter vinelandii* formed a different kind of nitrogenase when it was starved of molybdenum. It is in practice very difficult to starve a microbial population of molybdenum, because traces of the element are ubiquitous in dust, and for a few years the concept of an alternative non-Mo-nitrogenase was not widely accepted. Its reality was ultimately proved by genetic experiments: clear evidence was obtained that mutants of *A. vinelandii* that had had their normal genes for nitrogenase deleted (see section 6.2) still fixed nitrogen. Similar experiments in the UK demonstrated an alternative nitrogenase in *A. chroococcum* and established that it contained the metal vanadium in place of molybdenum. In due course vanadium was confirmed in the second nitrogenase of *A. vinelandii* and, to everyone's surprise, when the structural genes for the vanadium enzyme were also deleted yet another nitrogenase was

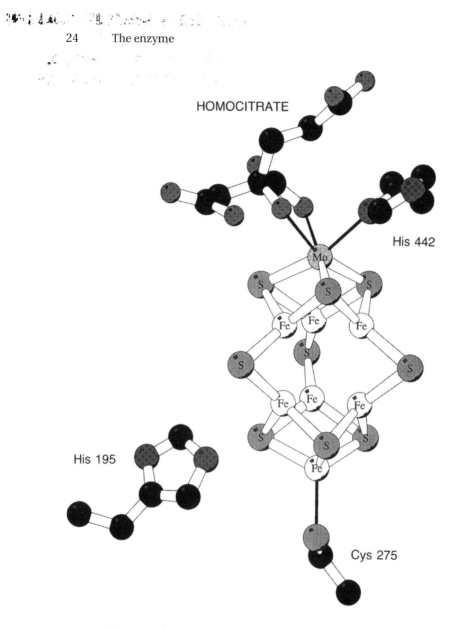

HOMOCITRATE

His 442

His 195

Cys 275

Fig. 2.3. The structure of FeMoco as revealed by X-ray crystallography of dinitrogenase. His and Cys stand for the amino acids histidine and cysteine and the numbers refer to their positions in the peptide chain of the protein. (Produced using the MOLSCRIPT program: Kraulis, P.J., 1991. MOLSCRIPT: a program to produce both detailed and schematic plots of protein structure. *Journal of Applied Crystallography.* **24**: 946–950, courtesy of Professor B.E. Smith.)

demonstrated. The third nitrogenase seemed to use an iron atom in place of Mo or V. It was absent from *A. chroococcum* but has since been found in the photosynthetic bacteria *Rhodospirillum rubrum* and *Rhodobacter* (earlier *Rhodopseudomonas*) *capsulatus*; the latter lacks the V-nitrogenase.

The alternative nitrogenases have not been as extensively studied as has molybo-nitrogenase, if only because they are relatively recent discoveries. However, it is clear that they have much in common with the conventional system both functionally and structurally. Both consume ATP, generate hydrogen when functioning, and reduce analogues of dinitrogen such as acetylene or cyanides. A quick indicator of the presence of an alternative nitrogenase is that some ethane appears along with ethylene when acetylene is reduced. Alternative nitrogenases are generally less efficient in ordinary circumstances than Mo-nitrogenase and are only formed in conditions of Mo-deprivation; they have been detected in clostridia and cyanobacteria as well as in the azotobacters and photosynthetic bacteria already mentioned.

As far as structure is concerned, like conventional nitrogenase, they consist of two proteins, one large and heteromeric, the other smaller, homomeric, and very like the regular Fe-protein. Both protein components of both alternative nitrogenases are very sensitive to oxygen. Their amino acid sequences show some, often considerable, mutual homology, and resemble their analogues in molybdenum nitrogenase. However, the larger proteins of both differ from that of the molybdenum nitrogenase in being hexameric: they are made up of six subunits, as three pairs of peptides (they are conventionally represented as having an $\alpha_2\beta_2\delta_2$ structure). The δ-subunit is quite a small ferredoxin-like peptide. The vanado-protein of *A. chroococcum* contains a functional equivalent of FeMoco: a 'FeVco' preparation has been obtained from the extracted vanado-protein and shown to activate a *nifB* mutant of *Klebsiella*. Unsurprisingly, activation was not as effective as with FeMoco, but the hybrid enzyme formed some ethane from acetylene, just as vanadium nitrogenase does. Similar tests on a *nifB* mutant with preparations from the larger alternative nitrogenase protein of *R. rubrum* suggest that it possesses a comparable 'FeFeco'. The structures of these co-factors has yet to be established.

2.6   An abberrant nitrogenase?

In 1997 a remarkable new kind of nitrogenase was discovered by Dr Meyer and his colleagues at Bayreuth University in Germany. They found it in a highly specialized species of bacteria called *Streptomyces thermoautotrophicus*. This microbe is autotrophic: like a plant it grows by reducing CO_2 to organic matter, but instead of utilizing sunlight it uses carbon monoxide (CO) as the reductant, obtaining the requisite energy by oxidizing it in air to CO_2. It can use hydrogen in place of CO, oxidizing it to water. On top of this exotic metabolism, it is also thermophilic: it grows best at 65 °C. It was isolated in 1992 from a rather special habitat: the warm overlay of a charcoal-burning heap in the Black Forest, where CO and CO_2 would have been abundant and air supplies adequate. At the time of its isolation it was reported to be able to fix modest but definite amounts of nitrogen, which occasioned some surprise because CO is a powerful inhibitor of conventional nitrogenase, but the extraordinary properties of its nitrogenase were not obvious until some 5 years later when preparations had been extracted and examined. Those extracts had rather low specific activities, but they undoubtedly converted $^{15}N_2$ to $^{15}NH_3$, using an enzyme system which appears to be unlike the types of nitrogenases discussed so far in several important ways. Naturally, it differs in not being inhibited by CO. It also does not reduce acetylene – indeed, acetylene has no effect upon it – and it is not at all sensitive to oxygen. It resembles conventional nitrogenase in that it evolves hydrogen when fixing nitrogen, and also in that it consumes ATP (as the Mg derivative), but MgADP does not interfere with the reaction, and less ATP is consumed per molecule of ammonia formed than is needed by conventional nitrogenases.

The new nitrogenase, too, consists of two proteins. One is a hetero-trimeric molybdoprotein ($\alpha\beta\gamma$ structure) which is probably involved in binding N_2 and contains Fe, Mo and S; it is only about half the molecular weight of the molybdoprotein of conventional Mo-nitrogenase. Moreover, in so far as it is known, its amino acid sequence is substantially different. The second protein is quite unlike conventional proteins too: a homomeric (α_2) protein (molecular weight about 400 kDa) which contains manganese, not iron, and which reacts with oxygen radicals (superoxides) generated during the organism's metabolism of CO. It passes electrons from superoxide radicals to the molybdoprotein. The system seems especially well adapted to the unusual mode of life of *S. thermoautotrophicus*; further details of the properties, structure and distribution this new nitrogenase, and any relationship to conventional Mo-nitrogenase, are eagerly awaited at the time of writing (late 1997).

2.7   Nomenclature

The name 'nitrogenase' is convenient but scientifically imprecise. Research has now established the existence of a family of enzymes, which can be distinguished as 'Mo-nitrogenase', 'V-nitrogenase' and 'nitrogenase-3', the last most probably an Fe-nitrogenase. Within these groups, the large, heteromeric protein is correctly termed 'dinitrogenase', because it actually binds the N_2 molecule, and its smaller, homomeric partner becomes 'dinitrogenase reductase', because it is the proximal electron donor to dinitrogenase. Mutant nitrogenases have appropriately modified names. In the case of the new nitrogenase described in section 2.6, the larger, heteromeric, protein would seem to be the dinitrogenase and the Mn-protein its dinitrogenase reductase. However, for most of this book the term 'nitrogenase' will continue to be used, signifying all types of N_2-fixing enzyme complexes.

3 Physiology

3.1 The need to exclude oxygen

The biochemical properties of nitrogenase present the organism with a number of problems of a physiological character. The overriding one arises from the sensitivity of the proteins themselves to oxygen: how, one can ask, do microbes fix nitrogen on a planet covered by 20% oxygen, when their enzyme is irreversibly destroyed on contact with that gas? The answer has proved very revealing. Nitrogen-fixing systems do not in fact function in the presence of oxygen, and the microbes which can fix nitrogen in air use various stratagems to exclude oxygen from the nitrogen-fixing site. When the organism is capable of anaerobic metabolism the stratagem is rarely complex, but obligate aerobic nitrogen-fixing bacteria have a more subtle problem because oxygen is essential for their general cell metabolism, without which no nitrogen could be fixed. So they are obliged to admit sufficient oxygen to sustain multiplication and ATP generation while not damaging nitrogenase. An even more awkward problem is faced by the cyanobacteria which are obliged to reconcile nitrogen fixation with the oxygen evolution step of photosynthesis. In this section the physiological processes whereby microbes protect nitrogenase from oxygen will be discussed starting with the simplest.

3.1.1 Avoidance

Nitrogen-fixing bacteria such as *Clostridium pasteurianum* or *Desulfovibrio gigas* (the sulphate-reducing bacteria, see section 4.1) are obligate anaerobes even when they are provided with fixed nitrogen, and they can be said to avoid the problem as far as possible. *D. gigas* shows no nitrogenase activity if a nitrogen-fixing population is exposed to air before testing but activity recovers in the absence of air. *C. pasteurianum* only grows in the absence of air, but the live cells can be centrifuged and otherwise handled in air without loss of nitrogenase; only when the cells are disrupted by some treatment is the nitrogenase destroyed by air. Though these anaerobes avoid oxygen, they can still, to a moderate extent, protect their nitrogenase from air.

3.1.2   Facultative nitrogen-fixing bacteria

Many bacteria are facultative anaerobes: capable of either aerobic or anaerobic growth. If such bacteria are also diazotrophic, they are usually so only in anaerobic conditions. Typical of this physiological type are nitrogen-fixing members of the genera *Klebsiella, Enterobacter, Bacillus* and photosynthetic bacteria such as *Rhodospirillum rubrum*.

The oxygen relations of *Klebsiella pneumoniae* have been studied in some detail and the statement that it cannot fix nitrogen in the presence of air requires qualification. Provided the population of *Klebsiella* is sufficiently dense, as it is, for example, inside a colony on agar, some fixation of nitrogen can occur. The determining factor is whether the cells, by their collective respiration, can lower the oxygen tension in their own vicinity to zero. Healthy colonies which fix nitrogen can be grown, for example, on agar slopes if the medium is free of inorganic nitrogen but contains a limiting amount of organic nitrogen. Continuous cultures of *Klebsiella pneumoniae* can be established in a liquid variant of such a medium and will grow and fix nitrogen indefinitely under an atmosphere containing oxygen provided the culture is not stirred too vigorously (fast stirring increases the rate at which oxygen dissolves in water). In such conditions, the organisms are actually growing aerobically: they are 'oxygen-limited' in the sense that they are consuming oxygen as fast as it is supplied. As will be explained in section 3.3, they are able to use products of their aerobic metabolism for nitrogen fixation. Physiologically, they are on the threshold of being nitrogen-fixing aerobes.

If detectable dissolved oxygen *is* present in the culture fluid, *K. pneumoniae* does not fix nitrogen. A simple question thus arises: does it make nitrogenase only to have it destroyed by oxygen? Or does it cease to make the enzyme when oxygen is present? In fact it does the second. Oxygen-damaged nitrogenase in cells can be detected quite easily by immunology or by electrophoresis of extracts of the organisms, and experiments using both techniques indicate that no nitrogenase is formed in the presence of oxygen. Oxygen thus regulates biosynthesis of nitrogenase in *K. pneumoniae*; the organism has some sensing equipment which recognizes the presence of free oxygen and 'switches off' nitrogenase synthesis. The nature of the primary oxygen sensor is not certain, but the effective switch is known to be a protein, the product of a gene called *nifL* (see section 6.3).

3.1.3   Microaerophily

There exists a class of bacteria which is naturally sensitive to oxygen yet which can grow aerobically. An example is *Lactobacillus casei*, one of the normal inhabitants of milk, which grows in air but only if the partial pressure of oxygen is low. Such bacteria are called microaerophiles; they grow microaerobically, to use the correct technical term, and are common in soil and specialized habitats to which air has limited access. Many of the aerobic diazotrophs behave like microaerophiles when fixing nitrogen in air, though in the absence of air they behave as perfectly normal aerobes. A well-established example occurs with the tropical microbe *Derxia gummosa*. When a population of these bacteria is spread on nitrogen-deficient nutrient agar, a few of the organisms form massive, glutinous colonies which, the acetylene test shows, fix nitrogen vigorously. Those colonies have a yellowish tint. The remainder are small, white colonies which fix no nitrogen and which do not develop further (Fig. 3.1). This colonial dimorphism, as it is called, is very consistent and sub-cultures from either small or large colonies show a similar pattern, It can be altered either by incubating the plates at a low oxygen tension (10% O_2 instead of the 20% O_2 of ordinary air) or by adding a small amount of an ammonium salt to the agar. The explanation of the dimorphic growth in air seems to be that media set with agar contain small amounts of fixed nitrogen (probably as an impurity in the agar itself, which is a polysaccharide obtained from seaweed) and a few colonies 'scavenge' this and reach a critical size such that, at their centres, the oxygen tension is low enough for nitrogen fixation to be initiated. These become the large, glutinous, yellowish ones; the other colonies obtain insufficient fixed nitrogen to reach a critical mass and do not develop. A lowered oxygen tension decreases the 'critical mass', so that all colonies get started; a trace of ammonium salt provides sufficient scavengeable fixed nitrogen for them all to reach the critical mass.

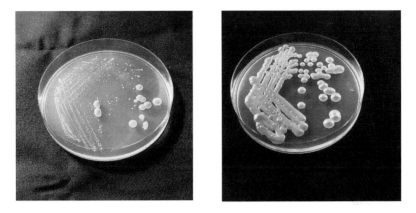

Fig. 3.1.   *Derxia gummosa* colonies grown on a jellified medium free of fixed nitrogen (*a*) in air, and (*b*) under nitrogen gas with only 5% oxygen. (Courtesy of Dr Susan Hill.)

A special (and, in its time, revolutionary) example of oxygen sensitivity is the demonstration that some strains of rhizobia show microaerobic diazotrophy. Rhizobia grow readily in air in media containing fixed nitrogen such as glutamate or nitrate but, for many years, scientists believed that rhizobia could only grow and fix nitrogen symbiotically; that is to say, in intimate association with a plant. However, in 1975 scientists in Australia and North America discovered that, at very low oxygen tensions, rhizobia belonging to a slow-growing genus called *Bradyrhizobium* would grow and fix nitrogen in the complete absence of plant material; this subject is discussed further in section 4.3.

Oxygen sensitivity amounting to microaerophily is shown by several other types of nitrogen-fixing aerobe, including *Xanthobacter flavus*, *Azospirillum lipoferum*, *Xanthobacter autotrophicus* and the methane-oxidizing bacteria. They are discussed further in section 4.3.

3.1.4   Respiration

The facultative and microaerophilic diazotrophs fix nitrogen in air only if their respiration enables them to decrease the oxygen level of their environment to innocuous levels. In all aerobic living things, respiration is a means of generating biological energy (as ATP) from the energy of

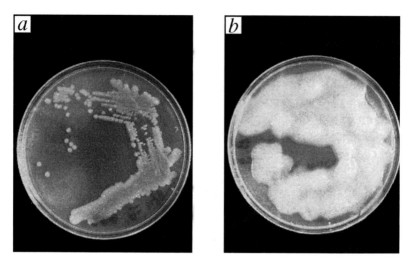

Fig. 3.2.   Colonies of *Azotobacter chroococcum* (left) and *Beijerinckia indica* (right) growing on a jellified medium free of fixed nitrogen. Note the glutinous quality of *B. indica* colonies.

combustion of foodstuffs; it is a highly efficient means of energy generation compared with the fermentative processes which are characteristic of most anaerobes. In these diazotrophs, respiration is performing a protective function: it protects the nitrogenase from oxygen damage. Certain of the azotobacters (such as *Azotobacter chroococcum*, see Fig. 3.2, or *A. vinelandii*) are very well adapted to aerobic nitrogen fixation and, though they do show oxygen sensitivity at high aeration levels or with hyperbaric oxygen, they have no difficulty in fixing nitrogen in air. A characteristic feature of such organisms is that they have exceptionally high respiration rates; 10 to 50 times the rate of *X. flavus*, for example. In azotobacters the protection of nitrogenase has become the dominant function of respiration: the organisms' respiration scavenges oxygen from the dinitrogen-fixing site. This process is known as 'respiratory protection' and it is now well established in the context of aerobic nitrogen fixation. (Similar processes may well operate in the protection of other oxygen-sensitive enzymes in microbes, but they are less clearly established.) *A. chroococcum* is particularly versatile in adjusting its respiration rate to the oxygen supply and a growing culture can double its respiration rate over about 20 minutes if the oxygen content of the atmosphere is doubled.

An obvious consequence of respiratory protection is that enormous amounts of carbon source must be consumed to scavenge the oxygen, so growth must be very inefficient in terms of the food consumed by the bacteria. This in fact proves to be true. An average laboratory culture of *A. chroococcum* consumes about 1 g of sugar (for example) to fix 10 to 15 mg of N. If the population is starved of oxygen, so that little respiratory protection is necessary, up to 45 mg N can be fixed for each gram of sugar consumed. Conversely, if aeration rates are high, fixation becomes very inefficient and amounts lower than 1 mg N are fixed per gram of sugar. Respiratory protection consumes a lot of substrate and, in nature, substrates are rarely available in great quantities.

A second consequence of respiratory protection concerns the generation of ATP. Although protection of nitrogenase is the dominant physiological function of *Azotobacter*'s normal aerobic respiration, the organism still needs biological energy and ATP must be generated. But respiratory protection might thus lead to generation of a vast excess of ATP. In practice, the organism proves to be able to regulate its ATP production at least to some extent. Though it is unable to uncouple respiration from ATP generation completely, it possesses two distinct biochemical respiration pathways. One is of high efficiency, and is dominant when the organism is growing with a relatively restricted supply of oxygen; the other is of lower efficiency, generating about a third as much ATP for a given amount of substrate, and is dominant in conditions of high oxygenation. Logically enough, the low efficiency pathway is used when respiratory protection is of primary importance.

Respiratory protection has evolved to a relatively elaborate level in *Azotobacter*, and one can see a clear sequence of increasing physiological sophistication when one considers oxygen-scavenging in the facultative bacteria, the microaerophiles and *Azotobacter*. But respiratory protection is not the only oxygen-restricting process in *Azotobacter*; a back-up protection apparatus will be discussed in section 3.1.6.

3.1.5   Slime

Aerobic nitrogen-fixing bacteria very frequently form large quantities of extra-cellular polysaccharides: the glutinous quality of colonies of *Beijerinckia*, for example, can be quite spectacular (Fig. 3.2). It has been suggested that this slime layer impedes diffusion of oxygen to the bacteria within and represents a crude physical barrier against excessive oxygenation, but gummy strains of *Klebsiella* or *Bradyrhizobium* seem to be just as oxygen-sensitive as non-gummy strains.

3.1.6   'Conformational protection'

Crude cell-free extracts of *Azotobacter chroococcum* or *A. vinelandii* do not contain the nitrogenase protein as an oxygen-sensitive solution. As was mentioned in section 2.1, the proteins are in an aggregate which can be sedimented in an ultra-centrifuge and within which the nitrogenase is relatively stable in air. It seems that, in the oxygen-tolerant aggregate, other proteins physically protect the oxygen-sensitive site from oxygen damage. This protected state reflects a physiological process in the living cell which was known in the late 1960s as 'conformational protection': the nitrogenase proteins were seemingly able to assume, reversibly, a conformation relative to protective molecules in which they were inaccessible to oxygen. They were then enzymically inoperative but undamaged. The concept explained a protective stratagem which *Azotobacter* had been shown to bring about in response to an oxygen stress too great for respiratory protection to cope with: in such circumstances the population ceased fixing nitrogen abruptly, but the nitrogenase remained undamaged and, provided exposure to oxygen stress had been brief, the cells resumed fixation as soon as more appropriate conditions of oxygenation were restored. The mechanism of this oxygen-dependent 'switch-off, switch-on' response was elucidated in Holland in the late 1970s: a low molecular weight protein containing iron and sulphur atoms is present in the aggregate which, given magnesium ions, binds in oxidising conditions to nitrogenase proteins and protects them from oxygen damage; it dissociates from nitrogenase in reducing conditions (Fig. 3.3). Switch-off, switch-on responses to oxygen stress have been found in other diazotrophs, including some anaerobes, suggesting that comparable intracellular physical protection of the enzyme system is widespread.

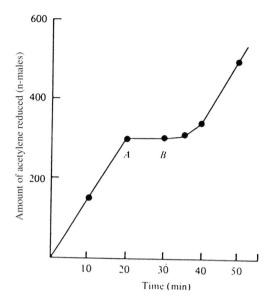

Fig. 3.3. 'Switch-on' and 'switch-off' of nitrogenase in *Azotobacter chroococcum* in response to aeration. Live *A. chroococcum* were shaken gently and their nitrogen-fixing activity measured using the acetylene test. Shaking, and therefore aeration, was intensified at *A* and returned to original value at *B*. Note the delay in reaching maximum activity again.

3.1.7   Heterocysts

Compartmentation, restriction of nitrogenase to special locations, is well established among cyanobacteria. Certain of the filamentous cyanobacteria such as *Anabena cylindrica* have, when they are fixing nitrogen, specialized cells called heterocysts at regular intervals along their filaments (Fig. 3.4). These do not appear when the bacteria are provided with adequate fixed nitrogen as ammonia (nitrate is less efficient in suppressing heterocyst formation). Substantial evidence has now accumulated to the effect that nitrogenase is normally restricted to the heterocysts and is largely if not completely absent from the vegetative filament. Correspondingly, the ordinary photosynthetic apparatus, which fixes CO_2 and evolves O_2, is present in the vegetative cell but the oxygen-evolving component is absent from the heterocysts. Thus the

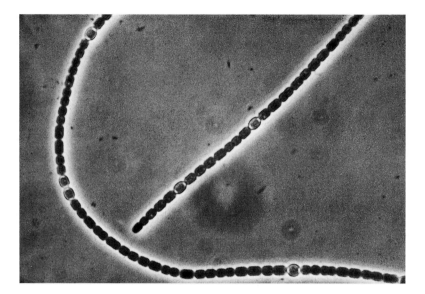

Fig. 3.4.   A photomicrograph of *Anabena cylindrica* showing two enlarged cells (heterocysts) on the principal filaments. (Courtesy of Dr P. Fay.)

heterocyst is a compartment from which the photo-generation of O_2 is excluded. Cyanobacteria are normally aerobic (or microaerophilic; some can show highly developed anaerobic metabolism which will not concern us here). Therefore they generally need some oxygen for respiration and air can very likely diffuse into the heterocyst to some extent: presumably respiratory protection can cope with that.

There exist filamentous cyanobacteria such as *Plectonema* which can fix nitrogen but which do not possess heterocysts. Characteristically they are very sensitive to oxygen, and to illumination (which leads to high oxygenation as a result of photosynthesis), and fix nitrogen only if the oxygen tension and illumination are low. *Anabena cylindrica* thus represents a phototrophic microbe whose nitrogen fixation processes are better adapted to the aerobic way of life than those of *Plectonema*. Some species of *Anabena* have the best of both worlds: their filaments fix nitrogen without benefit of heterocysts if the oxygen tension is very low.

3.1.8   Nodules

The root nodules of leguminous plants are gall-like growths within which the bacteroid forms of *Rhizobium* fix nitrogen (see Chapter 5). Nodules which are good at fixing nitrogen are called 'effective', and effectiveness has been known since the 1940s to be associated with the presence of a red pigment in the nodules. This pigment, which renders nodules slightly pink and which can be seen very clearly when nodules are sliced in half, is a haemoprotein rather like the haemoglobin of mammalian blood. It is called 'leghaemoglobin'. Its importance remained an enigma until the mid 1970s when workers in Australia established its function. It proves to transport oxygen, just like haemoglobin, but its affinity for oxygen is so high that it delivers it to the rhizobia at a concentration which is harmless to their nitrogenase. Thus leghaemoglobin represents a sort of oxygen buffer: a molecule which, by combining reversibly with oxygen, prevents the accumulation of high concentrations of free oxygen while at the same time providing the oxygen necessary for metabolism of *Rhizobium*. In addition, the nodules seem to possess a barrier which slows down diffusion of oxygen into them. The legume root nodule, then, is a highly specialized compartment in which nitrogen fixation and oxygen consumption are rendered physiologically compatible.

3.1.9   Other oxygen-restricting processes

The cyanobacterium *Gloeotheca* is coccoid. It does not form filaments nor does it possess heterocysts. Yet it is surprisingly oxygen-tolerant: it can actually fix N_2 under pure O_2 in the laboratory. Under the electron microscope it shows a complex intracellular membrane structure, which may imply subcellular compartmentation, but some scientists have speculated that it separates photosynthesis in time: that it fixes nitrogen when young, when its photosynthesis is still developing towards maximum O_2 production, and ceases fixation in a later period of maximal photosynthesis. There is some evidence that the pelagic marine cyanobacterium *Trichodesmium* forms mobile trichomes (rod-like microbes) which can aggregate in bundles if not too disturbed by the sea. Within these bundles, the central cells seem to lose photosynthetic

activity and gain nitrogen-fixing activity so the cluster as a whole becomes a nitrogen-fixing microcosm in which photosynthesis takes place
on the outside and nitrogen is fixed inside. Methane-oxidizing bacteria
attack methane with an oxygenase, an enzyme which introduces oxygen
into methane; some strains are diazotrophic and their property of consuming oxygen while oxidizing methane is thought to assist protection
of their nitrogenase. *Frankia*, the diazotrophic symbiont of actinorhizal
plants (see section 5.2) can fix N_2 either within or outside nodules; in
either case the diazotrophic forms often have vesicles which are thought
to assist O_2 exclusion by acting as diffusion barriers.

3.2   The biological reductant

Biochemists have been lucky in that nitrogenase interacts with sodium
dithionite (see section 2.2) and then performs its natural functions.
Sodium dithionite is not present in living cells; reducing power for the
enzyme is provided in living cells by specialized proteins called ferredoxins or flavodoxins. These are relatively small molecules, molecular
weights from 6000 to 24 000, which have in common the property of
existing in oxidized and reduced forms which can easily be converted
into each other. The reduced forms are strong reducing agents, capable
of reducing many other biological molecules (given an appropriate enzyme to catalyse the reaction). The reduced forms usually react rapidly
with air, becoming oxidized. Both types of protein can, in fact, exist in
more than two oxidation and reduction states, but only two of these are
important physiologically.

The ferredoxins are involved in a variety of biological processes, including photosynthesis in plants and pyruvate metabolism in anaerobic
bacteria. Those concerned in nitrogen fixation belong to the 'four iron'
class, which means that they have at least one cluster of four iron and
four sulphur atoms in the molecule and it is a curious fact that the many
iron atoms of nitrogenase (see section 2.1, Table 2.1) are in the form of
similar iron–sulphur clusters. Changes in the valency of the iron atoms
in the clusters are responsible for the special oxidation-reduction properties of the ferredoxins. It is not possible to discover precisely which
iron atoms are involved because the whole cluster behaves like an
oxido-reducible unit, but the kind of spectroscopy called electron para-

magnetic resonance (see section 2.3) indicates that the ferredoxins of aerobic nitrogen-fixing bacteria such as *Azotobacter* behave slightly differently from those of anaerobes such as *Clostridium pasteurianum*.

In *C. pasteurianum*, ferredoxin is the actual protein which reacts with nitrogenase and provides the reducing power for the conversion of N_2 to NH_3. It is probably the 'pyruvic phosphoroclastic' system, an enzyme complex responsible for pyruvate metabolism, which reduces the ferredoxin, so the link between diazotrophy and general cell metabolism is fairly clear in this organism. In the heterocystous Cyanobacteria, too, a ferredoxin is the primary electron donor to nitrogenase, and light, through the photosynthetic apparatus, can serve to generate electrons and reduce the ferredoxin. In *Klebsiella pneumoniae* it is a flavodoxin which is the primary reductant to nitrogenase and the flavodoxin is also reduced by a pyruvate-utilizing enzyme. In *Azotobacter chroococcum*, although a ferredoxin is present, the primary electron donor is thought to be a flavodoxin. *Azotobacter* flavodoxin is a yellow protein which, when half-reduced, is a striking blue colour (the 'semiquinone' form). When fully reduced it is colourless. In the living cells, the flavodoxin probably cycles between the reduced ('quinol') and semiquinone form. The semiquinone form of *Azotobacter* flavodoxin is unusually stable to oxidation by air, which may be why this protein rather than a ferredoxin is particularly suitable to *Azobacter*'s aerobic way of life.

Flavodoxins do not contain iron atoms; their oxido-reducible centre is a yellow, fluorescent molecule called a flavin.

3.3   The need for ATP

Nitrogenase requires biological energy as ATP in order to function (section 2.2). The source of this ATP is the normal metabolism of the microbe: catabolism of food leads to formation of ATP from ADP by a variety of chemical mechanisms, many of which are broadly understood but outside the scope of this chapter, and in cyanobacteria photophosphorylation, part of the process of photosynthesis, contributes ATP. If the ATP were not used for nitrogen fixation, it would be used for growth and multiplication. It follows that cultures of microbes which are fixing nitrogen are less efficient at converting food into cell material than are cultures which are using ammonia as their source of fixed nitrogen.

One can study the efficiency with which microbes convert their food into cell material by growing them in an environment in which the extent of their multiplication is limited by the amount of food supplied. For example, a culture of *Klebsiella aerogenes* grown in optimal conditions at 37 °C will convert 1 g of glucose into about 0.5 g (dried weight) of bacteria. Optimal conditions imply good aeration as well as growth in a continuous culture apparatus rather than the conventional laboratory batch culture; they also imply that the supply of glucose alone should limit the population density in the culture, not the supply of other nutrients such as phosphate, ammonium or potassium ions. With glucose, particularly when it is used by an anaerobic microbe, the number of molecules of ATP generated for each molecule of glucose consumed is often known as hence, quite easily, one can know the number of ATP molecules consumed to yield a gram of bacteria. Details of the experiments and calculations are not necessary here; the important principle is that yield studies on continuous cultures of microbes limited by a single carbon source can give information on the efficiency with which ATP, generated with the aid of that source, is utilized for multiplication. If one compared carbon-limited populations of a diazotroph which are using ammonium ions as their nitrogen source with similar populations which are fixing nitrogen, one finds that the yields with the latter are lower – sometimes substantially lower – and it is a reasonable assumption that most of this difference arises because ATP is diverted from biomass production to nitrogenase function.

Such experiments, although imprecise, give some interesting results. With *Clostridium pasteurianum* the yield difference corresponds to about 20 molecules of ATP diverted to reducing one N_2 molecule to two NH_3 molecules (the cell-free enzyme uses 16 ATP molecules; see section 2.2). With *Klebsiella pneumoniae*, some 30 ATP molecules are diverted to fixation. Section 3.1 described how nitrogen-fixing *K. pneumoniae* can be grown with air provided its respiration is adequate to keep the dissolved oxygen near to zero. In those circumstances the yield increases, but this seems to be because the microbe can generate more ATP per molecule of glucose when provided with a little air; it does not become more efficient in its ATP economy for nitrogen fixation.

It is less easy to do comparable experiments with the aerobe *Azotobacter* for three reasons. One is general: aerobes are very much more efficient at generating ATP so the number of ATP molecules produced per gram of cells is difficult to calculate. The second arises because *Azotobacter* can alter the efficiency with which it generates ATP in response to the oxygen tension (see section 3.1.4). The third, and more drastic, is that respiratory protection leads to consumption of substrate connected neither with biosynthesis nor with nitrogenase function. Despite these difficulties it is possible to make approximations and extrapolations which provide an estimate of the minimum efficiency of *Azotobacter*, and the surprising result is that this organism has the most efficient ATP economy of all. Something in the region of six ATP molecules become diverted into nitrogenase function: the living organisms appear to be twice as economical as the cell-free enzyme and over three times as economical as the anaerobic nitrogen-fixing bacteria. Azotobacters not only have the most effective oxygen-excluding processes (sections 3.1.4 and 3.1.7), they also seem to be exceptional in their ATP economy. They seem to be among the best adjusted, free-living, nitrogen-fixing micro-organisms for the terrestrial environment as it exists today.

It is, of course, impossible to do such experiments with nodules, but analyses of C and N balances in leguminous plants suggest that nodule bacteria are quite as efficient as azotobacters, if not more so.

3.4   The need to control hydrogen evolution

I mentioned in section 2.3 that nitrogenase produces a minimum of one molecule H_2 for every N_2 fixed. This side-reaction occurs in the living diazotroph; *Klebsiella* and *Clostridium pasteurianum* evolve hydrogen naturally during anaerobic growth and, if they are simultaneously fixing nitrogen, a proportion of the hydrogen comes by way of nitrogenase. The side-reaction, yielding H_2, may account for the low ATP efficiency of nitrogen-fixing cells of these organisms, since the ATP used for this process is effectively wasted. The enzyme responsible for normal hydrogen evolution (i.e. that formed in the normal metabolism of these bacteria, not via nitrogenase) is called hydrogenase, and it can catalyse both the uptake and evolution of hydrogen. *Azotobacter* evolves no hydrogen in normal circumstances and it contains a kind of hydrogenase which is not reversible: it catalyses only the uptake of hydrogen. This enzyme can be selectively inhibited by certain chemicals and, when this is done,

Table 3.1. *Acetylene and carbon monoxide cause* Azotobacter *to evolve hydrogen. Cultures of nitrogen-fixing* A. chroococcum *were incubated at 30 °C in closed flasks with a large air space to which the gases shown were added. Hydrogen was measure in the air space by gas chromatography*

Added gas	nmol of H_2 evolved/hour
None	None
10% acetylene (C_2H_2)	20
5% carbon monoxide (CO)	440
10% C_2H_2 + 5% CO	760

Table 3.2. *Formation of hydrogen by symbiotic nitrogen-fixing associations. Nodules were cut from the roots of leguminous nitrogen-fixing plants (see Chapter 5) and the amount of hydrogen formed measured in relation to their nitrogen-fixing activity*

Plant	Nitrogenase activity diverted to hydrogen evolution
Soya bean	35%
Garden pea	25%
Cowpea	< 1%

Azotobacter cells start evolving hydrogen from their nitrogenase. This is illustrated in Table 3.1; it is the sort of evidence which has led to the view that, though nitrogenase does evolve some hydrogen in the living cell, hydrogen is trapped by hydrogenase in *Azotobacter* and recycled. It can be used to generate ATP and/or reducing power. The use of recycled hydrogen to generate some extra ATP may contribute to the unusually efficient ATP economy of *Azotobacter* mentioned earlier.

Cyanobacteria and photosynthetic bacteria show photo-evolution of hydrogen which is partially attributable to nitrogenase. A fascinating development was the discovery that some symbiotic nodule bacteria show a similar side-reaction. In some plant–bacteria associations as much as 35% of their available ATP and reducing power can be lost in H_2 evolution (Table 3.2). In terms of the productivity of crops such as soya beans, this is a finding of some agricultural importance: a 'tight' (non-

H_2-evolving) symbiosis obviously fixes more nitrogen per unit of solar energy than a 'loose' (H_2-evolving) one. The strain of symbiotic bacterium rather than the plant cultivar determines the tightness of the association in this respect because some strains of *Bradyrhizobium japonicum*, the soya bean symbiont, can make hydrogenase and others cannot. The non-leguminous associations, as far as they have been examined, are tighter than most of the leguminous associations.

3.5   The need to assimilate molybdenum or vanadium

That nitrogen-fixing systems (bacteria and nodulated legumes) need extra molybdenum (Mo), usually provided as sodium molybdate, has been known since the 1930s. In some agricultural regions (e.g. parts of Australia) the soil contains rather little Mo and the use of a fertilizer containing molybdate, together with a legume such as clover, considerably upgraded such land in the 1940s. The reasons for the effect of Mo is clear: it is part of the prosthetic site of nitrogenase. In the natural environment, Mo is ubiquitous but rarely abundant; diazotrophs possess physiological means of taking it up from the environment, a process which is partly understood, and usually have means of storing it, in association with special Mo-storage proteins. Vanadium has long been known to decrease the need for Mo, at least in azotobacters, a fact which makes sense with the discovery of a vanadoprotein as part of an 'alternative' nitrogenase (see section 2.4). Vanadium is also ubiquitous but generally scarce; the fact that synthesis of V-nitrogenase is repressed by Mo suggests that it provides azotobacters with a physiological 'back-up' nitrogen-fixing system for use in case of shortage of Mo.

3.6   The need to regulate nitrogenase activity and synthesis

Nitrogenase is an expensive enzyme in terms of biological energy: it is slow (so diazotrophs need a lot of it) and it consumes a lot of ATP even in the relatively efficient *Azotobacter*. Therefore, diazotrophs have developed very precise means of regulating both its activity and its synthesis.

Details of the regulation of the activity of nitrogenase are not wholly understood, but in outline the position seems simple. ADP, the product

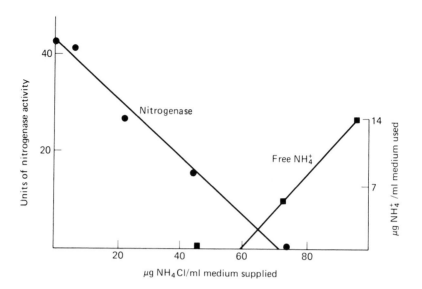

Fig. 3.5.   Repression of nitrogenase activity by ammonia. *Azotobacter chroococcum* was grown continuously in liquid media with increasing amounts of ammonium chloride (NH$_4$Cl). The nitrogenase activity (tested with acetylene) of the populations so obtained is given in arbitrary units. The amount of ammonium ion remaining in the medium after removal of the grown bacteria is also plotted; none was detected unless nitrogenase activity was completely repressed.

of utilization of ATP by nitrogenase, inhibits the enzyme. ATP is essential for the enzyme to work at all, so the ratio of ATP to ADP in the neighbourhood of the enzyme can determine the rate at which the enzyme works. In a test tube this 'control' process works convincingly; in the living cell the situation is very probably more complex – if only because the ratio of adenine nucleotide concentrations (ATP, ADP and AMP) has a controlling effect on the activity of many enzymes other than nitrogenase, and separate controls of the latter simply by the ratio of two of these nucleotides would thus be difficult. In several diazotrophs – *Bacillus, Rhodospirillum, Desulfovibrio* and *Azotobacter*, but not *Klebsiella* – addition of ammonia to the population causes immediate 'switching-off' of nitrogenase activity, a process which presumably conserves ATP. The mechanism involves the attachment of an ADP-ribosyl group to the smaller of the two nitrogenase proteins to turn the enzyme off, and its release to restore activity; quite how ammonia triggers these changes is still obscure. Nitrogenase activity in *Azotobacter* is regulated by oxygen in that partially conformationally protected populations arise at partially inhibitory oxygen tensions.

Much is now known about regulation of the synthesis of nitrogenase at the molecular genetic level, but detailed discussion will be postponed to section 6.3. At a descriptive level, ammonia (the primary product of nitrogenase activity) suppresses synthesis of the enzyme: if the environment contains sufficient ammonia for the microbe's needs, it does not make any nitrogenase. If there is less than sufficient, but still some, the microbe makes only sufficient nitrogenase to satisfy its requirement for fixed nitrogen. The situation is illustrated in Fig. 3.5. Other nitrogen sources – nitrates, urea, amino acids – often suppress nitrogenase synthesis equally well, probably because they are converted transiently to ammonia during metabolism of the cell. By growing organisms with minimal ammonia and letting them use it up, one can study rates of nitrogenase synthesis and learn that, in *Azotobacter* and *Klebsiella*, synthesis of both nitrogenase proteins is regulated together. When ammonia is absent, a 'message' is made in the cells telling its biosynthesis machinery to make the enzyme; this 'message' has a half-life of about 20 minutes.

Azotobacter is good at protecting its enzyme from oxygen, but *Klebsiella* is less effective in this way. It would be biologically improvident for *Klebsiella* to make nitrogenase in air, only to have it destroyed by oxygen and, as was mentioned in section 3.1.2, oxygen can repress nitrogenase synthesis by *Klebsiella*: neither protein is formed at all when detectable free oxygen is present in the environment. An unexpected finding is that, despite its tolerance of oxygen, the same principle applies to *Azotobacter*: oxygen, albeit at quite high concentrations (about 20 times the oxygen tension needed to cause 95% inhibition of activity) causes repression of nitrogenase synthesis.

4   The free-living microbes

The advent of the acetylene test provided scientists with a rapid and generally very effective tool for the assessment of the diazotrophic potential of organisms, symbiotic systems and environments. New organisms and associations were discovered and, inevitably, some older ones came under suspicion and, in some cases, were rejected. The process of revision is continuing in the 1990s. In no case, yet, has a positive acetylene test given a false result *when applied to a living system* (a few inanimate chemical mixtures will reduce acetylene to ethylene in water; examples are colloidal palladium under H_2, or a mixture of cysteine, sodium molybdate and sodium borohydride). The converse proposition, that all biological nitrogen-fixing systems give a positive acetylene test is so nearly true as to be reliable, but exceptions (discussed in section 4.3) are known and, as indicated in Chapter 3, adverse environmental conditions, particularly high oxygenation, can obscure potential diazotrophy in many microbes.

This chapter is concerned with the organisms which fix nitrogen alone and which do not normally, or necessarily, have a symbiotic partner such as a plant. Table 4.1 lists many of the currently accepted organisms in this group; several will reappear in the next chapter because they can in fact form symbiotic associations; indeed, some are best known as symbionts. The list has some important features. In the first place all are prokaryotes; the list includes some methanogenic bacteria which belong to the recently distinguished Domain Archaea (below), but no nucleate organisms are present at all. Even very simple eukaryotes such as yeasts, single-celled algae and fungi are absent (before 1970, some yeasts and nucleate microfungi were thought to be capable of fixing nitrogen – see section 4.5). The second feature is that, although only about a third complete, it has a reasonably comprehensive span and indicates that, among the many thousands of species of bacteria that have so far been named, diazotrophy is scattered widely among diverse physiological types but is nevertheless rare. In fact the property seems to be restricted

Table 4.1. *List of genera of microbes which include nitrogen-fixing species or strains*

	Genus or type	Species (examples only)
Strict anaerobes	*Clostridium*	*C. pasteurianum[a], C. butyricum*
	Desulfovibrio	*D. vulgaris, D. desulfuricans*
	Methanosarcina[d]	*M. barkeri*
	Methanococcus[d]	*M. thermolithotrophicus[e]*
Facultative	*Klebsiella*	*K. pneumoniae, K. oxytoca*
(aerobic when	*Bacillus*	*B. polymyxa, B. macerans*
not fixing	*Enterobacter*	*E. agglomerans (Erwinia herbicola)[b]*
nitrogen)	*Citrobacter*	*C. freundii*
	Escherichia	*E. intermedia*
	Propionibacterium	*P. shermanii, P. petersonii*
Microaerophiles	*Xanthobacter*	*X. flavus[a], X. autotrophicus*
(normal aerobes)	*Thiobacillus*	*T. ferro-oxidans*
when not fixing	*Azospirillum*	*A. lipoferum[a], A. braziliensis[a]*
nitrogen)	*Aquaspirillum*	*A. perigrinum[a], A. fascicilus[a]*
	Methylosinus	*M. trichosporum*
	Bradyrhizobium[c]	*B. japonicum*
	Herbaspirillum	*H. seropedicae[a]*
	Burkholderia	*B. brasilense[a]*
Aerobes	*Azotobacter*	*A. chroococcum[a], A. vinelandii[a]*
	Azotococcus	*A. agilis[a]*
	Azomonas	*A. macrocytogenes[a]*
	Beijerinckia	*B. indica[a], B. fluminis[a]*
	Derxia	*D. gummosa[a]*
Phototrophs	*Chromatium*	*C. vinosum*
(anaerobes)	*Chlorobium*	*C. limicola*
	Thiopedia	
	Ectothiospira	*E. shapovnikovii*
Phototrophs	*Rhodospirillum*	*R. rubrum*
(facultative)	*Rhodopseudomonas*	*R. palustris*
Phototrophs	*Plectonema*	*P. boryanum*
(microaerophiles)	*Lyngbya*	*L. aestuarii*
	Oscillatoria	
	Spirulina	
Phototrophs	*Anabena*	*A. cylindrica, A. inaequalis*
(aerobic)	*Nostoc*	*N. muscorum*
	Calothrix	
	(7 other genera	
	of heterocystous	
	blue-green algae)	
	Gloeotheca	*G. alpicola*

[a] Signifies that all reported strains of the species fix nitrogen. For commentary, see text. [b] Synonyms. [c] See Table 5.2. [d] An archaeon, see section 4.1. [e] A thermophile.

to between 100 and 200 prokaryotic species. The third point is a little more subtle. Obviously ability to fix nitrogen is universal among genera and species for which diazotrophy is a criterion of the definition of the group: a naturally occurring non-diazotrophic *Azotobacter* would not be recognized as such. However, it is not necessarily universal among groups for which diazotrophy is not definitive, such as the genus *Clostridium* or the species *Klebsiella pneumoniae* and *Bacillus polymyxa*; thus only some 30% of *K. pneumoniae* isolates exhibit diazotrophy, and some 80% of *B. polymyxa* strains. Doubtless many new species of diazotroph remain to be discovered, but in principle it seems that diazotrophy only occurs in bacteria, and even among them the property appears to be rare and haphazardly distributed.

It is important to be clear that free-living nitrogen-fixing bacteria, at least when in the free-living state, are rarely of great importance in the terrestrial nitrogen economy. They need too much organic matter. The only free-living organisms of major agronomic importance are the cyanobacteria which, being capable of photosynthesis, are not limited by the availability of carbon substrates in the soil (these matters are discussed in more detail in section 4.4). Despite their generally rather low agronomic significance, the free-living nitrogen-fixing bacteria are of exceptional scientific importance and, since the majority of them can form productive associations of one kind or another, knowledge of their general microbiology is very valuable.

Table 4.1 includes one thermophilic microbe. Thermophiles are specialized bacteria, which grow at temperatures well above 44 to 50 °C, which is the upper limit for normal 'mesophilic' microbes. They are found in hot springs, fermenting compost heaps, geothermal waters, hot water systems and (surprisingly) ordinary soil; often they grow only at elevated temperatures. For many years the existence of thermophilic diazotrophs was widely doubted, but now at least three strains of thermophilic diazotrophs are well established: one is a cyanobacterium (*Mastigocladus*) from a hot spring at 60 °C; another is the archaeon shown in Table 4.1, *Methanococcus thermolithotrophicus*, which fixes at 64 °C; the third is *Streptomyces thermoautotrophicus*, scoring 65 °C – see section 2.6.

4.1   The anaerobic bacteria

The first nitrogen-fixing microbe to be discovered was *Clostridium pasteurianum*, (Fig. 4.1) obtained by one of the father-figures of microbiology, S. Winogradsky, in Paris in 1893. He called it *C. pastorianum*, meaning a spindle-shaped organism (*clostridium*) from the field, but its species name became transformed to *C. pasteurianum* (meaning pertaining to another great microbiologist, Louis Pasteur) in the 1920s. It metabolizes carbon sources such as glucose to butyric acid $+CO_2+H_2$, with minor products and is related to other butyrate-forming clostridia such as *C. butyricum* and *C. aceto-butyricum*. While by no means universal, the ability to fix nitrogen seems not uncommon among the clostridia and they are widespread in soils, in decaying vegetable matter (e.g. compost heaps) and even in the rumina of ruminant animals such as cows or sheep (they do not fix nitrogen there, and are probably only passengers, ingested with the food). They are obligate anaerobes; though sometimes tolerant of traces of oxygen, their metabolism can make no use of oxygen. In adverse conditions they form spores which resist heating and drying and these, no doubt, account for the ease with which *C. pasteurianum* can be isolated from almost any soil or water sample.

A more exacting group of anaerobes, which includes nitrogen-fixing strains, is the sulphate-reducing bacteria (e.g. *Desulfovibrio*). These are organisms which use the oxygen atoms of sulphate ($SO_4{}^{2-}$) for respiration, producing sulphide and water, A typical overall metabolic reaction is:

$$2CH_3.CHOH.CO_2Na + CaSO_4 \rightarrow 2CH_3CO_2Na + CaS + 2H_2O + 2CO_2$$

| sodium lactate | calcium sulphate | sodium acetate | calcium sulphide | [4.1] |

In most conditions the sulphide hydrolyses to the gas hydrogen sulphide, contributing to the repellent smell of highly polluted environments. Though most free-living diazotrophs are not of great environmental importance, *Desulfovibrio* does have a special ecological rôle in the sea because it is probably the commonest indigenous marine nitrogen-fixing genus which is not phototrophic (i.e. does not require light). It is considered to be the principal contributor of biologically fixed nitrogen to marine sediments. Other nitrogen-fixing bacteria, including butyric clostridia, klebsiellae and even azotobacters, can sometimes be found in offshore waters, but they are thought to be terrestrial bacteria washed into the sea with land water and are probably not truly marine nitrogen fixers.

An important group of anaerobes was added to the diazotrophs in 1984: the methanogens. These are very fastidious anaerobes which form methane in muds, aqueous sediments and the rumina of ruminant mammals – even in our own intestines. Although examples were suspected of diazotrophy as long as 40 years ago, definitive proof was held up because they are very difficult to obtain and cultivate in pure cultures. Their special interest arises from the fact that, although classified as bacteria, they belong to the third Domain of living things: the Archaea (earlier Archaebacteria). This is a category of creatures (all non-nucleate microbes) which diverged from both bacteria and the nucleate ancestors of higher organisms (plants and animals) at an early stage in the evolution of terrestrial life, so it is interesting to discover diazotrophy (and, in one strain, thermophilic diazotrophy) in these microbes but not in higher forms of life.

Certain photosynthetic anaerobes fix nitrogen and are mentioned in section 4.4.

4.2   The facultative bacteria

These are a physiological group of bacteria which are able to grow either with or without the oxygen of air when provided with fixed nitrogen, but which can only fix nitrogen anaerobically. It took many years before their ability to fix nitrogen became firmly established but it now seems that they are very common diazotrophs in soils and waters.

A member of this group, *Klebsiella pneumoniae*, has appeared frequently in earlier chapters. Numerous strains and species of *Klebsiella* have been examined for nitrogen fixation and it is a curious fact that the property is distributed over about 30% of them with no particular correlation with the source of the strain: a *Klebsiella* from the gut is as likely – or as unlikely – to fix nitrogen as one from a soil or water sample.

Klebsiella belongs to a group of bacteria called 'coliforms', so-called because they resemble in appearance a bacterium called *Escherichia coli* normally present in the guts of animals, including man. No fully authenticated natural strains of diazotrophic *E. coli* have been isolated, though laboratory strains able to fix nitrogen have been constructed genetically (see Chapter 6), but the species *E. intermedia* can fix nitrogen. The coliform genus *Citrobacter*, found in soils, foods and the intestines of termites, includes nitrogen-fixing strains; so does the genus *Enterobacter* (which also inhabits intestines, as well as soils and waters). Some coliform plant commensals such as the species *Erwinia herbicola*, include nitrogen-fixing strains.

The genus *Bacillus*, is a second major group of facultative bacteria which includes diazotrophs. *Bacillus polymyxa* and *Bacillus macerans* are two species among which nitrogen fixation seems common (but again not entirely universal). The bacilli are characterized by their ability to form spores and thus survive well in environments subject to desiccation or periodic heating. Their behaviour has something in common with clostridia and, indeed, the borderline between air-sensitive bacilli and aerotolerant clostridia is difficult to define. The propionic bacteria mentioned in Table 4.1 are reported by Russian workers to fix nitrogen anaerobically. They occur in fermenting milk and their possession of this property is rather surprising.

Certain phototrophic bacteria are facultative anaerobes and only fix nitrogen anaerobically; they are discussed in section 4.4.

4.3   The aerobes

The best-known aerobic nitrogen-fixing bacteria are members of the genus *Azotobacter*; indeed, *A. chroococcum* (Figs. 4.1 and 3.2) was the second free-living nitrogen-fixing microbe to be discovered after *C. pasteurianum*. It was reported by a famous Dutch microbiologist, M. W. Beijerinck, in 1901. Today the '*Azotobacter* group' comprises several genera: *Azotobacter*, *Azomonas*, *Azotococcus*, *Beijerinckia* and *Derxia*; they are rather alike in appearance and physiological character and none is capable of growth without air. They also look rather alike: large granular microbes, ovoid or rod-shaped, sometimes motile, which form spreading, glutinous colonies on nitrogen-free agar media. In general, the azotobacters are well adjusted to oxygen: they fix nitrogen in ordinary air relatively efficiently. Some, such as *Derxia*, have difficulty in initiating growth in air (see section 3.1.3) and within the group there is in

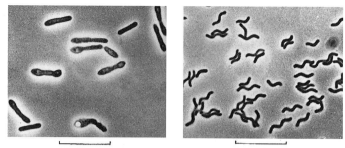

10 μ | 10 μ
Clostridium pasteurianum | Rhodospirillum rubrum

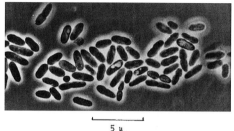

5 μ
Azotobacter chroococcum

Fig. 4.1.   Photomicrographs of selected diazotrophic bacteria.
C. pasteurianum shows sporulating forms. (Courtesy of Dr Crawford
Dow.)

fact a range of oxygen-sensitivities which may reflect the extent of their
adjustment to an aerobic way of life. If *Derxia* is most sensitive to oxygen
stress, then *Beijerinckia* is intermediate and *A. chroococcum* and *A.
vinelandii* are least oxygen-sensitive. Even these, however, can be inhibited by really high oxygenation.

Most aerobic diazotrophs are less well adapted to air than the
azotobacters and were recognized only relatively recently. The reason is
that, when they fix nitrogen, they behave like microaerophiles. The
microaerophilic character of *Xanthobacter flavus* (named for the yellow
colour of its colonies) was mentioned in section 3.1.3; two or three
bacteria of rather similar properties were described along with it when it
was discovered by scientists in the former USSR in 1961. These organisms were thought to be mycobacteria, relatives to the causative organisms of tuberculosis and leprosy, but they were probably misclassified.

X. flavus has a lot in common with a group of aerobic nitrogen-fixing bacteria called *Xanthobacter* (earlier *Corynebacterium*) *autotrophicus*, many strains of which can grow chemotrophically (see section 1.5) with gaseous hydrogen as an energy source, deriving energy from its oxidation to water. They, too, are very sensitive to oxygen. It is interesting that strains of *Bradyrhizobium japonicum* which possess hydrogenase (section 3.4) can grow chemotrophically with H_2, thus resembling *X. autotrophicus* in this property as well as in oxygen sensitivity.

Another broad group of aerobes which behave like microaerophiles when fixing nitrogen is the spirilla, represented by the species of *Aquaspirillum* and *Azospirillum* listed in Table 4.1; azospirilla, along with several other genera of microaerophilic diazotrophs, are involved in symbiotic associations with grasses (see section 5.4). Azospirilla are generally rather large, curved or corkscrew-like bacteria, which often grow as a pellicle some way below the surface of a culture medium (where, presumably, the dissolved oxygen tension is just right). Reports of diazotrophic spirilla had been in the scientific literature for many years but their oxygen sensitivity made their ability to fix nitrogen difficult to establish with certainty. Similar considerations apply to *Thiobacillus ferro-oxidans*, a remarkable microbe which grows in acid mine waters (pH 2 to 4) and obtains energy by oxidizing dissolved ferrous ions to ferric ions. Like the nitrifying bacteria (section 1.5), this organism is a chemotroph: the energy of iron ion oxidation enables it to convert CO_2 to organic matter and to multiply. Strains of this resourceful microbe can fix nitrogen, but only at low dissolved oxygen tensions.

The methane-oxidizing bacteria were also long suspected of being able to fix nitrogen – partly because the soil around blow-holes of natural gas (methane) is generally more fertile and richer in fixed N than elsewhere. Many strains can, indeed, fix nitrogen and they show great sensitivity to oxygen, but authentication of this was delayed for a different reason. They provide, in fact, an instance in which the normally reliable acetylene test for diazotrophy breaks down. Acetylene inhibits the enzyme system responsible for methane oxidation so, if these bacteria are grown with methane and tested with acetylene, the result is negative: they cannot generate any ATP from methane and so cannot use what nitrogenase they may have. All strains can use methanol (methyl alcohol) in place of methane so, if this substrate is provided, their possession of nitrogenase shows itself clearly. Another instance in which the acetylene test can fail also involves methane metabolism: the methanogens (section 4.1) are very sensitive to acetylene and unusually small amounts must be used for the acetylene test or their diazotrophy will be missed.

Perhaps the most dramatic example of microaerobic fixation occurs among the rhizobia, the symbiotic partners of leguminous plants. For many decades a dogma existed to the effect that rhizobia never fix nitrogen away from a host plant. In the mid 1970s this view was overthrown when scientists in Australia and Canada successfully grew certain slow-growing types of rhizobia such as 'cowpea' strains (see section 5.1) in media entirely free of plant material. The secret is that these microbes are very bad at respiratory protection (see section 3.1.4) and therefore their nitrogen fixation is highly oxygen-sensitive (they grow in air without difficulty given fixed N as glutamate or nitrate). Only with the most careful control of the dissolved oxygen concentration will they fix nitrogen in liquid media; on solid (jellified) media things are a little easier provided some fixed nitrogen is around, because colonies of bacteria grow and within these at least some of the microbes find the oxygen tension just right for fixing nitrogen. Though there is now no doubt that these bacteria can make nitrogenase and fix nitrogen outside a host plant, they have never grown in media entirely free of fixed nitrogen: they need a small amount of ammonia, glutamine or some other fixed nitrogen source to start them off. Whether this is a subtle reflection of their extreme oxygen sensitivity or a special physiological requirement for fixation is not yet known.

4.4   The phototrophs

As explained in section 1.5, phototrophs are autotrophs which make use of light to fix CO_2 and multiply. There are phototrophic representatives of all the classes of microbes discussed in the preceding section of this chapter. The strict anaerobes such as *Chromatium* and *Chlorobium* are sulphur bacteria: they oxidize sulphides, elemental sulphur or thiosulphates to sulphates while absorbing light with the aid of chlorophyll and carotenoid pigments (as do green plants). The energy therefrom is used for growth and can be coupled to nitrogen fixation. They are coloured:

Chlorobium species are green, *Chromatium* and *Thiopedia* are red or purple. Probably many other types of coloured sulphur bacteria can fix nitrogen; all are obligate anaerobes. The coloured non-sulphur bacteria, of which *Rhodospirillum rubrum* (Fig. 4.1) and *Rhodobacter capsulatus* (both purple) are examples, grow phototrophically, but only anaerobically, and their photosynthetic processes do not lead to evolution of oxygen. They can fix nitrogen anaerobically in light. It is possible to grow these organisms heterotrophically (i.e. with organic substrates such as sugar or alcohols) and aerobically, but only in the dark. In these circumstances they only fix nitrogen in microaerobic conditions.

Anaerobic and facultative phototrophic bacteria may be of minor importance to the terrestrial nitrogen cycle except in highly specialized environments such as sulphur springs or polluted waters, but by far the most important phototrophs from an ecological viewpoint are the cyanobacteria. They differ from the coloured bacteria in being able to grow phototrophically in air (though some can actually mimic the sulphur bacteria and grow phototrophically and anaerobically with sulphide). They also have a more sophisticated photosynthetic equipment which, like that of higher plants, evolves oxygen as a product of photosynthesis. Thus they face the problem, discussed in section 3.1.7, of excluding photo-generated oxygen from nitrogenase, and they can be grouped according to the effectiveness with which they do this. As was described in that section, the most sophisticated have compartments called heterocysts to which nitrogenase is restricted. Eight genera of heterocystous blue-green algae are known and they all fix nitrogen readily in air at normal levels of illumination. *Anabena cylindrica* has been studied in depth in several laboratories and its properties are probably typical. The heterocysts are formed only in nitrogen-deficient environments (see Fig. 3.4); they appear at regular intervals along the filament and the nitrogenase activity of the population is proportional to the heterocyst count. Nitrogenase has been extracted and partly purified from heterocysts; their lack of photosystem II (responsible for the photo-evolution of O_2 was also mentioned in section 3.1.7. The evidence that heterocysts are oxygen-restricting compartments for nitrogenase is overwhelming.

The genus *Gloeotheca* (earlier *Gloeocapsa*) was mentioned in section 3.1.9: it is a unicellular diazotrophic cyanobacterium which may have an unusual oxygen-restricting apparatus.

Several genera of non-heterocystous cyanobacteria exist and, characteristically, they are microaerophilic when fixing nitrogen. They also dislike high illumination levels. They are frequently found in association with other microbes, in shaded or polluted waters, and they are clearly less sophisticated diazotrophs than their heterocystous cousins.

As stated earlier, the cyanobacteria are ecologically by far the most important free-living diazotrophs. The relative unimportance of free-living nitrogen-fixing bacteria in the terrestrial nitrogen economy was emphasized at the opening of this chapter. For reasons which range from inefficient ATP production and/or utilization to massive respiratory protection, the ordinary diazotrophs rarely find sufficient energy-providing substrates in nature for them to upgrade the nitrogen status of the environment significantly. The cyanobacteria are in a different position because they fix nitrogen by using solar energy and, provided they are not in the dark, energy ceases to be the critical limitation. Though their major ecological contribution to the nitrogen cycle is made symbiotically (see Chapter 5), they probably contribute, as free-living organisms, some hundred times as much N to a soil as do bacteria such as *Clostridium pasteurianum* or *Azotobacter chroococcum*. An estimate made by Russian workers of 30 kg N/ha per year (cf. 0.1 to 0.3 kg for *Azotobacter* or *C. pasteurianum*) for their contribution to soils is probably an underestimate because it was made before fixation by the non-heterocystous cyanobacteria was discovered. Cyanobacteria are the first living things to colonize arid, ice-bound and devastated areas when conditions begin to improve; they pioneer biological colonization of rocks, buildings, newly exposed sand, chalk and other infertile substrate; they often colonize living and decaying plant surfaces such as tree-trunks. They even fix nitrogen at the surface of ordinary soils, in puddles after rainy weather. They are probably the only important aerobic diazotrophs in the open sea, as well as on rocks and coral reefs.

4.5   The 'ghosts'

Several groups of microbes which were once thought to fix nitrogen are now known not to do so. Sometimes called 'ghosts' by microbiologists, they include yeasts and some other simple fungi, some species of the genus *Pseudomonas*, *Azotomonas* and several species less clearly identi-

Fig. 4.2.   Colonies of a 'ghost' on nitrogen-free agar. They bear a
superficial resemblance to nitrogen-fixing bacteria (Fig. 3.2) but consist
largely of mucoid material and very few microbes. (Courtesy of Professor
H.J. Evans.)

fied. They all appear to grow successfully on (or in) nitrogen-free media,
and they often form thick, glutinous colonies like those of an authentic
nitrogen fixer such as *Beijerinckia*. Fig. 4.2 is a close-up photograph of
such colonies. Often the N-content of the environment is augmented
after growth of such 'ghosts'. They appear for two principal reasons.

First, some yeasts and pseudomonads are exceptionally good at utilizing traces of fixed nitrogen present as contamination in bacteriological
reagents or in ordinary air. Town air contains oxides of nitrogen and free
ammonia; a colony on a Petri dish culture can scavenge these gases,
converting the nitrogen oxides to nitrate or nitrite, assimilating these,
thus facilitating further diffusion of such gases towards itself. When this
happens, the bacteria are seen under the microscope to be very sparse in
the colony – most of it is mucilage. The N-content of an ordinary
microbe is 12 to 15% of its dry weight, but such 'ghost' bacteria can have
as little as 3 to 6% N. In one case known to me, a yeast 'ghost' had only
1% N and gave a most convincing impersonation of a diazotroph.

Second, a less frequent reason for a 'ghost' is contamination. *Clos-*

tridium pasteurianum cannot grow in air but it can multiply inside colonies of some other bacteria such as *Pseudomonas*. Instances are known in which putative diazotrophic aerobes were actually non-fixers harbouring undetected diazotrophic anaerobes within their colonies. Such associations are misleading to the scientist, but they may reflect an important ecological situation if the two types of microbe assist each other. But this topic is more appropriate to the next chapter.

5   The plant associations

The existence of symbiotic nitrogen-fixing systems, in which bacteria fix nitrogen in association with the roots of plants, was suspected by Sir Humphry Davy early in the nineteenth century, but was confirmed only in 1886–8 by the German scientists H. Hellreigel and H. Wilfarth. These systems are now known to be much the most important contributors to the nitrogen cycle, from both ecological and agricultural points of view. The associations formed by leguminous plants in which soil bacteria of the broad group called rhizobia (singular: rhizobium) colonize nodules on their roots (Fig. 5.1) are the best known and, since they include such crops as clover, lucerne (alfalfa) and pulses (peas and beans), they are traditionally the most important to agriculture. In recent years, however, new types of association have been discovered and some known ones have become recognized as being of unsuspected importance. A few have proved not to be real diazotrophic symbioses at all.

5.1   The legumes

The Leguminosae is a family of flowering plants found in both tempera-ture and tropical zones. Botanists believe that they were originally tropi-cal. They range from small but widespread plants such as clover, through flowering plants such as lupins and bushes such as brooms, to shrubs and trees (e.g. *Acacia*). The botanical properties of the Leguminosae will not be discussed here since they are available in any advanced botany textbook. The family comprises three major subfamilies, the Papilion-oidae (the largest group), the Mimosoidae and the Caesalpinoidae. Be-tween 80 and 90% of the species in the Papilionoidae form nodules, but only about a quarter of the Mimosoidae and relatively few of the Caesal-pinoidae. Well over 12 000 species of Leguminosae are known, the ma-jority of which fix nitrogen; it is a sobering thought that less than 50 have been exploited for agricultural purposes and of these about seven are

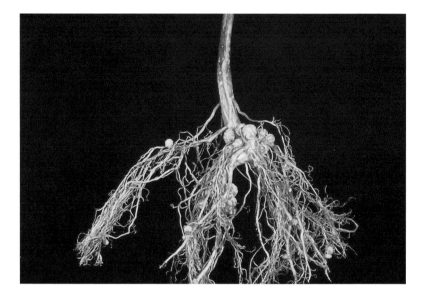

Fig. 5.1.   Nodules on a soya bean root. (Courtesy of Professor
H. J. Evans.)

regularly used in agriculture. Table 5.1 lists a selection of familiar or
agriculturally important members of the family Leguminosae. In vir-
tually all nodulated legumes the nodules are on the roots, but in the
tropical weed *Sesbania* they occur on the stem. Nodules may be rounded
in appearance (called determinate) as in Fig. 5.1 or lobed (called indeter-
minate); these shapes reflect more subtle differences in their origin and
development.

Until 1973, scientists believed that the Leguminosae were the sole
hosts of rhizobia but in that year an instance was discovered in which
cowpea rhizobia nodulate a non-leguminous tropical plant now known
to be a species of *Parasponia*. It is at present the only non-leguminous
genus known to be colonized by rhizobia. Both leguminous plants and
species of rhizobia show a degree of specificity: certain types of plant are
colonized only by certain types of bacteria. A classification of rhizobia
into 'cross-inoculation' groups on these lines is given in Table 5.2, but it
is not very rigid because the groups may overlap.

The classification of the rhizobia has undergone several alterations

Table 5.1. *Selected list of Leguminosae*

Papilionoidae (temperate)	
Pisum sativum	Garden pea
Vicia faba	Broad bean
Phaseolus vulgaris	French or kidney bean
Trifolium repens	White clover
Medicago sativa	Lucerne
Ulex	Gorse
Lupinus polyphyllus	Garden lupin
Lotus corniculatus	Bird's foot trefoil
Melilotus officinalis	Melilot
Papilionoidae (tropical and subtropical)	
Glycine max	Soya bean
Archis hypogaea	Ground nut (peanut)
Cicer arietinum	Chick-pea
Mimosoidae	
Acacia	Includes the 'wattles'
Xilia dolabriformis	Ironwood of India
Mimosa pudica	Sensitive mimosa
Caesalpinoidae	
Cassis fistula	Senna
Tamarindus indica	Tamarind
Ceris siliquastrum	Judas tree

during the past two decades. Microbiologists used to place all rhizobia in one genus, *Rhizobium*, which could be subdivided broadly into fast-growing and slow-growing types as in Table 5.2. However, the differences in rates of multiplication which the two subdivisions showed seemed to reflect real microbiological differences between the species and in the 1980s a new genus, *Bradyrhizobium* was created to encompass the slow-growers. More recently an intermediate genus called *Sinorhizobium* was accepted, and the rhizobium which colonizes the stem nodules of *Sesbania* proved to be distinctive and is now classified in its own genus, *Azorhizobium*. Doubtless yet other genera will be named in

Fig. 5.2.   Nodules on the stem of the tropical weed *Sesbania*.
(Courtesy of Dr B. Dreyfus.)

future. The genus *Rhizobium* still encompasses most fast-growing types, but within that genus the species names have been revised. In earlier literature one finds the clover group referred to as *R. trifolii* and the *Phaseolus* (bean) group as *R. phaseoli*, but they are now regarded as varieties ('biovars') of the pea group, *R. leguminosarum*. In this book I shall use the systematic names for microbiologically defined types but shall continue with the general name rhizobium to cover all types.

The dogma that rhizobia never fix nitrogen except within the nodule of a plant was widely accepted until it was overthrown in the mid 1970s (see section 4.3). It crumbled in two stages. Tissue cultures of plant cells are not difficult to obtain and rhizobia were first shown to be able to fix

Table 5.2. *Cross-inoculation groups of rhizobia*

Organism	Examples of host plants	
Fast-growing types		
Alfalfa group	*R. meliloti*	Melilot, lucerne
Pea group	*R. leguminosarum*	Peas, lentils, some beans (*Vicia*)
Clover group	*R. leguminosarum bv. trifolii*	Clover
Phaseolus group	*R. leguminosarum bv. phaseoli*	Beans (*Phaseolus*)
Slow-growing types		
Lupin group	*B. lupini*	Lupins
Soybean	*B. japonicum*	Soya bean
Cowpea group	'Cowpea rhizobia'	Cowpea, peanuts, acacia

nitrogen in association with tissue cultures of a leguminous plant (soya bean). Then an active association with a tissue culture of a non-legume (tobacco plant) was demonstrated and, within months, conditions were discovered for fixation without plant material at all. As was described in section 4.3, slow-growing rhizobia prove to be highly sensitive micro-aerophiles when fixing nitrogen alone. Ability to fix nitrogen *ex planta*, as it is termed, is restricted to the slow-growing class of which the 'cowpea' group is typical; fast-growing rhizobia such as *R. meliloti* have not so far been induced to fix nitrogen convincingly without plant material. Perhaps this is only a matter of time.

Despite the discovery of fixation *ex planta*, the really important process for biological productivity is conducted in symbiosis with a plant. Nearly all soils contain rhizobia, normally not fixing nitrogen, and these colonize appropriate plants at about the stage in the plant's growth season when leaves begin to appear. The process of colonization of host plants has been studied in detail for several decades, and a fascinating sequence of mutual interactions has been revealed involving the genes (see section 6.8) of both plant and microbe. In outline, the roots of the host plant secrete attractant molecules called flavonoids which invite the right species of rhizobium to enter into symbiosis. The bacteria sense the appropriate attractant and migrate towards the plant root, where the flavonoid actually initiates a genetic programme in the bacteria which sets them on course to becoming symbionts. One of their earliest responses is to produce their own signalling substances, complex sugar derivatives called lipochitooligosaccharides, which diffuse to nearby root hairs and cause them to curl inwards. This enables the bacteria to penetrate into the root by way of a so-called infection thread: a kind of tunnel which winds in among the cells of the root without actually infecting them. In due course special types of plant cell are actually invaded, the bacteria multiply within them and they become engorged. Nevertheless, the bacteria remain covered by a plant-derived coat called the peribacteroid membrane within which they cease multiplying, become mis-shapen and rich in nitrogenase, and initiate nitrogen fixation. Such semi-dormant cells are called bacteroids; the clusters of nitrogen-fixing bacteroids within their membrane actually resemble organelles such as plastids and have been called 'symbiosomes'. Meanwhile the plant has grown nodule tissue around them and surrounded them in a 'bath' of leghaemoglobin. In a space of 1 to 2 weeks, ideal conditions for symbiotic nitrogen fixation have thus been provided by the plant: the peribacteroid membrane isolates the bacteroids from the plant cytoplasm and contains the infection, but it allows certain fatty acids such as malic and succinic acids (but not sugars) together with minor essential nutrients to reach the bacteroids and be consumed. The membrane also releases to the plant the ammonia generated by bacteroid nitrogen fixation as soon as it is formed. And the whole structure ensures that the bacteroids receive, by way of leghaemoglobin, a steady supply of oxygen at a concentration too low to damage nitrogenase.

The plant can abort the process of colonization, or even prevent it, if adequate fixed nitrogen is available in the soil. The amount of nitrogen fixed by the bacteroids is regulated to match the requirements of the plant precisely: no leakage occurs unless the plant suffers trauma. Fixation of nitrogen continues until (usually) seed formation, when it ceases and the nodules become sensescent. However, some rhizobia remain viable and, as the nodule (or, in annuals, the plant) ages and dies, many survive in soil for the next season's growth.

Details of the colonization process differ in character and timing from plant to plant. Notably different from the rest are the stem nodules of *Sesbania*, which develop, in response to a relatively simple exchange of chemical signals between plant and bacterium, after *Azorhizobium* has entered the plant through dormant root primordia, which lie in vertical rows along the stem.

Generally speaking, there is clear evidence that plants recognize and admit only the correct species of rhizobium, but among strains of those species there can be considerable divergence in symbiotic effectiveness. Thus soya bean plants can be infected by a variety of strains of *Bradyrhizobium*, and these can differ in their ability to fix nitrogen efficiently. A really effective strain, as it is called, forms healthy-looking nodules with a distinct pink colour; if the nodules are cut open, the interior is distinctly red. Ineffective rhizobia form pallid, sometimes greenish, nodules. Once a plant has been colonized by a certain strain of rhizobium, others seem to be excluded, so it is important in agriculture to ensure that the rhizobia which infect a crop plant are effective. In many areas the natural rhizobia are relatively ineffective, and cultures of good, effective rhizobium strains damp-dried in peat are now available to farmers for 'pelleting' seeds: for coating them with the live, effective rhizobia. Even so, the local rhizobia can sometimes compete with the introduced ones and supplant them. Means for supplying live, effective rhizobia to crop plants can assume great agricultural importance, particularly in underdeveloped countries, and really reliable means for doing this have still to be developed.

One problem arises from the very effectiveness of the rhizobia themselves. Leguminous plants do not admit rhizobia if the soil is reasonably rich in fixed nitrogen. If a plantation of peas, for example, has successfully nodulated on a poor soil and given a good harvest, much of the non-harvestable material (roots, stems and leaves) will die, decay and add fixed nitrogen to the soil, so upgrading it. In the next year, nodulation will be discouraged by this available nitrogen, and the effective rhizobia will tend to die out compared with the natural strains. This 'second year death' can, in extreme cases, cause the soil to revert to its low status by the third year. The practice of agriculture involving the use of symbiotic systems is, therefore, different from that with fertilizers, and requires a different outlook on the part of the farmer. Other problems, such as the responses of the bacteria to herbicides, drying, freezing and so on, need consideration, too. On the other hand, the cost of fertilizer (both material and transport costs) becomes minimal and pollution problems resulting from the escape of fertilizer into wells, streams, rivers and so on are much diminished. For a discussion of the relative advantages and disadvantages of biologically versus chemically fixed nitrogen the reader should consult more specialized works. In summary, energy, transport and pollution costs weigh against chemical fertilizers but they can never be abandoned completely. This topic will be returned to in Chapter 7.

The importance of red colour in effective nodules is now understood. It is due to leghaemoglobin, the protein which supplies oxygen to the bacteroids but avoids oxygen-damage to their nitrogenase. The legume nodule is a complex biological structure for protecting rhizobial nitrogenase from oxygen. As I indicated earlier, leghaemoglobin is actually produced by the plant, not the bacteria, although the nitrogenase is produced by the bacteroids. The enzyme from *B. japonicum* bacteroids has been extracted and substantially purified; like the nitrogenases of free-living organisms it consists of two iron-containing proteins, one containing molybdenum in addition. It is oxygen-sensitive and consumes ATP; it reduces the usual substrates; it is so similar to the nitrogenases of free-living bacteria that its components will cross-react with those of *Azotobacter.*

5.2   The actinorhizal symbioses

The fact that a number of non-leguminous plants have root nodules has been known for many decades. With some, such as the subtropical conifer *Podocarpus*, the 'nodules' are associated with a symbiotic fungus (mycorrhiza; see section 5.5.2). Others may be gall-like products of microbial or viral infection. But some 120 species are now known, principally trees and shrubs, whose root nodules harbour diazotrophs which are not rhizobia. Table 5.3 lists some examples of such plants. The best known example is *Alnus* (alder); *Alnus* has a wide geographical distribution and often colonizes infertile or devastated areas (Fig. 5.3). One of the first to be recognized (in the late nineteenth century) was *Eleagnus*, but knowledge of the physiology of non-leguminous nitrogen fixation has progressed relatively slowly – even the ecological importance of these plants, now known to be considerable, has only been recognized in the past four decades. The slow progress in understanding the symbiosis arose in part because the plants are tough, woody and

Table 5.3. *Some actinorhizal plants*

Genus	Common name of an example
Alnus	Alder
Casuarina	Australian pine
Ceanothus	Californian lilac
Eleagnus	(*E. maculata*)
Myrica	Bog myrtle
Hippophae	Sea buckthorn
Purshia	Bitter brush
Dryas	(*D. octopetala*)

Fig. 5.3.   Seedling of alder showing root nodules.

often slow-growing (many inhabit unusual environments) but mainly because the symbiotic diazotroph proved extremely difficult to obtain in pure laboratory culture. However, its isolation was reported in 1978 from *Alnus* and a shrub called *Comptonia*. It belongs to a group of filamentous bacteria called Actinomycetes, hence the name 'actinorhizal' for these symbioses, and pure cultures, like crushed nodules, can be used to infect seedling plants. The microbe can travel in soil from one plant to another, probably by growing as a filament, like a fungus.

The organisms have the generic name *Frankia* and some species have been named on the basis of their infectivity to plants. For, like rhizobia, the frankias show 'cross-inoculation' groups: for example, microbes which nodulate *Alnus* are said not to nodulate *Eleagnus* or *Ceanothus*, though they will infect *Myrica*. But the organism from *Myrica* does not infect *Alnus*. However, as with rhizobia, there is not always agreement about the reliability of the information on which such beliefs are based.

When *Frankia* colonizes a plant such as *Alnus* it appears to induce root hair deformation, the microbe's hyphae penetrate the root at the fold and grow along a channel between the host's cells. In due course certain plant cells become infected and a nodule develops in which the filament can be seen together with swellings called vesicles. Haemoglobin-like proteins have also been detected in actinorhizal nodules but whether they act like the leghaemoglobin of legumes is not clear.

There is some evidence for restricted geographical distribution of the microbes. *Ceanothus* is regularly nodulated in the USA but rarely in Europe. *Alnus* is almost always nodulated in Europe but European *Myrica* by no means always has nodules.

The actinorhizal diazotrophic systems are of only modest agricultural importance today, partly because their potential value is not widely known. Yet their input of N on a global scale must be enormous: thousands of square kilometres of Washington State and Oregon, USA, are populated by *Alnus*, *Ceanothus* and *Purshia* together. *Alnus* alone can add 100 kg N/ha per year to soil just from leaf fall (about one third the input of a good legume such as lucerne) so the combined contribution must be impressive. Plants in this group have little agricultural potential as crops or forage, but alder could be valuable in forestry. Scrub alders have been used traditionally to upgrade poor soils in Asia and, of the trees used to stabilize dykes in Holland, every third one is customarily an alder. The exploitation of these symbioses obviously requires further attention.

Research on the biochemistry and physiology of the actinorhizal symbioses has also been relatively slow-moving. Cell-free preparations of nitrogenase have been obtained from *Alnus* nodules but little is known of the enzyme beyond that it shows the familiar oxygen sensitivity. As I mentioned in section 3.1.9, frankias can fix N_2 *ex planta* and then show relatively low sensitivity to oxygen. In such cultures vesicles resembling those in nodules can be seen.

5.3   The cyanobacterial symbioses

The ecological importance of cyanobacteria as free-living diazotrophs was emphasized in section 4.4. Numerous instances are known in which a cyanobacterium associates with a plant. It is curious that, whereas the rhizobia and *Frankia* species form associations only with the most advanced plants, the dicotyledenous angiosperms, the cyanobacteria tend to favour the more primitive members of the plant kingdom.

5.3.1   The lichens

The most primitive plant–microbe associations of this kind are the lichens, in which a diazotrophic cyanobacterium associates with a fungus (Fig. 5.4). Since it is a little difficult to judge which is the host and which is the symbiont, the terms 'phycobiont' and 'mycobiont' are used to describe the bacterial and fungal partner respectively. In fact, most lichens do not fix nitrogen: the phycobiont is a green alga. Rather less than 10% of known lichens have a cyanobacterium as phycobiont (often *Nostoc* or *Calothrix*: see Table 4.1). As would be expected, light is necessary for nitrogen fixation by lichens, but the organisms otherwise tolerate extreme conditions. In antarctic species, nitrogen fixation takes place at 0 °C and the organisms survive extended freezing; they also survive drying but do not fix nitrogen until they become damp again. They colonize the most unpromising environments: rocks, stones, tree-trunks, walls and roofs of buildings and they are principal vegetation of some arctic and tundra regions. As 'reindeer moss', large subarctic lichens are an important food source for reindeer.

Fig. 5.4. *Lobaria*, a forest lichen from northwestern USA, which includes diazotrophic cyanobacteria in loci called cephalodia. (Courtesy of Professor H. J. Evans.)

5.3.2   Liverworts

The liverworts include species (e.g. *Blasia pusilla*) which have cyanobacteria of the genus *Nostoc* associated with the underside of the thallus. There has been some argument about whether fixation of nitrogen by the *Nostoc* actually benefits the plant but work with the isotope of nitrogen $^{15}N_2$ seems to have confirmed modest incorporation of fixed nitrogen into the plant.

5.3.3   Pteridophytes

Azolla (Fig 5.5) is a tiny water fern which grows, often to a high density, in tropical waters. It is only 2–3 mm across and it harbours an *Anabena* (see Table 4.1) in a cavity at the base of the leaf-like frond. The microbe is particularly rich in heterocysts (see section 3.1.7) which presumably protect its nitrogenase from the considerable oxygenation which its own and the plant's combined photosynthesis can develop in the tropical

Fig. 5.5. *Azolla microphylla* plants from Paraguay. (Courtesy of Dr I. Watanabe.)

sunshine. The symbiotic *Anabena* is very difficult to grow away from its plant host, but this has been done successfully on occasions; the fern can grow quite well without the alga if fixed N is available. In nature, the association is very effective and holds the record for nitrogen fixation rates. *Azolla* is an important green manure for rice culture: in the high season, the rate of input of N into a rice paddy can exceed that of a good legume crop on land. Introduced into temperate zones, *Azolla* can add substantial amounts of N to lake waters. It is one of the most promising systems for the improvement of tropical and subtropical agriculture; it has also been used as pig fodder in Vietnam.

5.3.4   Gymnosperms

A number of species of this phylum, among the family of cycads, form nodule-like structures on the roots near, but beneath, the soil surface. One of the best described is a cycad called *Macrozamia* (Fig. 5.6); its roots develop club-shaped excrescences (called 'coralloid' roots, from their resemblance to coral) which include a deep green zone. These roots are

Fig. 5.6. *Macrozamia*, a cycad from Western Australia, showing
coralloid roots. (Courtesy of Professor M. J. Dilworth.)

well colonized by diazotrophic cyanobacteria such as *Nostoc* and they fix
nitrogen; the curious thing is that the process is not necessarily light-
dependent. Cycads grown in a glasshouse do fix more nitrogen if the roots
are illuminated but field samples do not necessarily show this effect.

5.3.5   Angiosperms

A single example is known of a higher plant which forms a cyanobacterial association. *Gunnera* is a subtropical plant which is invaded by a
species of *Nostoc*. As a result, nodules are found at the bases of the leaves
and within these nitrogen is fixed and made available to the plant. As

with *Azolla*, the *Nostoc* is particularly rich in heterocysts in the symbiotic state, possibly because its location is one to which photo-evolved oxygen has ready access.

5.4   Associative symbioses

Previously called diazotrophic biocoenoces, this is a name for a group of mutualistic systems in which there is some interdependence between the partners though both can grow satisfactorily apart. They involve grasses (Gramineae) principally, and the first one to become established scientifically was between a sand grass called *Paspalum notatum* (Fig. 5.7) and a species of *Azotobacter*, *A. paspali*. *Paspalum* grass is widespread and persistent in tropical and subtropical areas such as Brazil and Australia; it is of little value as a forage grass but some related grasses could be of value in animal production. Many cultivars of *Paspalum* form an association with *A. paspali* which is rather specific: *A. paspali* grows around the roots of the plant, though the dominant nitrogen-fixing microbes in the surrounding soil are usually different (*Beijerinckia* or *Derxia*, for example). In due course the azotobacters form a sheath, partly mucilaginous, over the roots and cease growing, though they do not cease fixing nitrogen. Those plants which are able to form this association show benefits in vigour, dry weight and nitrogen content. It is interesting that the discoverer of the association, Dr Johanna Döbereiner of Brazil, suspected its existence several years before she was able to confirm it with the acetylene test and with isotopic nitrogen. The reason is that the association is very sensitive to oxygen: if you dig up plants and test them as you might test legumes, with acetylene or isotopic nitrogen in air, the nitrogenase of *A. paspali* ceases functioning. Only if tests are performed with a low level of oxygen – probably closer to conditions down in the soil – are good positive results obtained. The input of N to soil by natural *Paspalum* associations is probably small, but the acetylene test indicates that some 20 kg N/h per year could potentially be brought to soil by this system.

The second such symbiosis to be discovered occurs with another tropical/subtropical grass called *Digitaria decumbens*. In this case the microbial partner is *Azospirillum lipoferum* (see Table 4.1), a bacterium which is widespread in both temperate and tropical soils. The spirilla do

Fig. 5.7. *Paspalum notatum* in an experimental plot in Brazil. The
cultivar in the left foreground forms an association with *Azotobacter
paspali*; the cultivar towards the centre, marked with a stake, does not
and has grown poorly. (Courtesy of Dr J. Döbereiner.)

not form a sheath around the roots of *Digitaria* but in fact they invade
the root tissue, forming a layer beneath the epidermis where they cease
multiplying but continue to fix nitrogen. There is no doubt that they
supply nitrogen to the plant, but the scale on which they do this is not
clear at the time of writing, and may normally be small. The *Azospirillum–Digitaria* association caused considerable excitement around 1975
because of reports that the microbe would associate with some varieties
of maize. The prospect of nitrogen-fixing cereal crops was thereby
opened, which, if real, would be of enormous benefit to world food
production. More sober reports indicate that the bacteria are indeed
catholic in their tastes: they will associate with some cultivars of maize,
sorghum or millet; however, their effect on the N-content of the plants is
small and often undetectable.

The prospect of finding bacteria which would form associative symbioses with rice, other cereals or graminaceous crops has provoked
considerable interest in those diazotrophic bacteria which inhabit the

rhizospheres of such plants. In recent years several new types have been reported, including species of the new genera *Herbaspirillum*, *Azoarcus* and *Burkholderia* as well as a diazotrophic species of the long-known genus *Acetobacter* called *A. diazotrophicus*. All are microaerophilic diazotrophs; most actually enter the host plant tissues, some multiply in the sap, and some appear to enter the plant cells. As with *Azotobacter paspali* and *Azospirillum lipoferum*, research on the effectiveness of these symbioses is complicated by the fact that they often form hormone-like products which stimulate plant growth without actually providing the host with significant amounts of fixed nitrogen, but in some instances promising nitrogen increments have been found. Recent research in Brazil suggests that a combination of *Acetobacter diazotrophicus* with *Herbaspirillum*, when associated with sugar cane, can give fixation rates comparable to a leguminous symbiosis.

5.5   Casual associations

There are a number of instances in which nitrogen-fixing bacteria associate with higher organisms and in which the benefit incurred by one partner in the association is small or negligible.

5.5.1   Casual root associations

About thirty species of grasses are now known which have diazotrophs loosely associated with their roots, including varieties of maize (one instance of which pre-dates the excitement about maize and *A. lipoferum*). They have been detected by the acetylene test but not studied in detail – some are 'couch grasses' found even in British gardens (e.g. *Cynodon dactylon*; Fig. 5.8). Free-living diazotrophs are often found round the roots of dicotyledenous plants too, no doubt living on root secretions and benefiting themselves rather than the plant. Some temperate weeds such as hedge woundwort and the persistent ground elder also give positive acetylene tests. In all these instances the presumption is that the association is casual (involving no lasting interdependence) but some may be true symbioses. Mycorrhiza, a group of symbiotic fungi, associates symbiotically with many types of plant, assisting uptake of nutrient (such as phosphates) and enhancing vigour in ways

Fig. 5.8.   The couch grass *Cynodon dactylon* or 'Bermuda grass'.

which are not often understood. In the mid 1960s there seemed to be some evidence that the mycorrhiza of a gymnosperm, *Podocarpus*, fixed nitrogen but more recent evidence suggests that the mycorrhiza simply supply a favourable environment for casual fixation by ordinary soil bacteria such as klebsiellae and bacilli. The borderline between casual associations and symbioses can be difficult to define: in principle, a symbiosis benefits both partners directly and a casual association benefits only one, but indirect benefit to the second partner can obviously arise in many associations which seem at first sight to be casual.

Though these seemingly loose root associations are at present of no obvious economic value – indeed, they can be a nuisance in both agriculture and horticulture – there is no doubt that they are of considerable ecological importance in the colonization by plants of harsh, relatively infertile or devastated land.

5.5.2   Leaf associations

The association of cyanobacteria with the bases of leaves (*Gunnera*) or leaf-like organs (*Azolla*, liverworts) are true symbioses. More casual associations occur when nitrogen-fixing bacteria grow on the leaves of plants, then become washed into soil by rain and thereby upgrade the nitrogen status of the soil near the plant roots. The surface zone of the leaf where fungi or bacteria able to utilize leaf exudates grow, is called the 'phyllosphere' (which is why these associations have been called 'phyllocoenoses'). In many plants of the wet tropics, such as cocoa and sugar cane, the phyllosphere is populated by *Beijerinckia* which very probably upgrade the local soil after tropical rainstorms. In other instances – fir trees in temperate regions, for example – it is very doubtful whether even a casual association really exists. Much of the research in this area has not taken account of the ease with which ghosts (section 4.5) can be isolated from such habitats.

A more sophisticated phyllocoenosis, which was thought for some years to be a diazotrophic, is the leaf nodule system. Certain tropical plants (*Ardesia*, *Psychotria* – the latter is quite a popular house plant in Britain) have tiny nodules on their leaves which, as has been known for several decades, are colonized by bacteria with consequent benefit to the plant. Among the bacteria readily isolated from such nodules are diazotrophic klebsiellae (see Table 4.1) so, in the early 1960s, the idea that the beneficial effect on the plant was due to fixation of nitrogen was plausible. Since then the idea has had to be discarded, for two main reasons: (1) the beneficial bacteria are by no means always diazotrophic; and (2) even if they are, the acetylene test indicates that they cease fixing nitrogen when they colonize the leaf nodule. It seems probable that the benefit to the plant arises from the formation by the bacteria of hormone-like substances in the leaf nodule, not from nitrogen fixation.

5.5.3   Destructive associations

By altering the nitrogen status of the local environment, free-living diazotrophs can enhance microbial degradation processes. They do so in compost heaps, decaying leaf litter and other natural environments where the C:N ratio is high. A clear-cut case occurs in the rotting of timber: nitrogen-fixing coliform bacteria (see section 4.2) can multiply alongside wood-rotting fungi and accelerate the rotting of wood by making fixed nitrogen available to the fungus. Here the bacteria benefit from being able to utilize the products of fungal degradation of cellulose.

5.5.4   Associations with animals

Termites and other wood-eating insects have diazotrophic bacteria including citrobacters and *Azoarcus* in their intestines and, when the host is on a low nitrogen diet, they actually fix some nitrogen. The acetylene test indicates that their contribution to the nitrogen status of the insect is, at best, small. Butyric clostridia (see section 4.1) are normally found in the rumina of sheep and cows but their nitrogen-fixing activity is slight and could not account for more than 1% of the animal's N turnover. They have probably been ingested with food; this is certainly the case with winter reindeer which, after feeding on reindeer moss, can have a lot of nitrogenase activity in the rumen due to the phycobiont (see section 5.3).

Klebsiellae capable of nitrogen fixation have been isolated from human intestines but, again, it is unlikely that they ever contribute substantial amounts of N to their host. Generally speaking, the animal intestinal tract contains sufficient free ammonia to repress (see section 3.6) any worthwhile nitrogen-fixing activity.

An exception seems to be the shipworm, a mollusc which bores into the hulls of wooden ships: this creature has cellulose-decomposing diazotrophic bacteria in a gland in its intestinal tract and can obtain up to a third of its N by way of these symbiotic bacteria.

6   Genetics and evolution

The study of the hereditary information which enables organisms to fix nitrogen is somewhat simplified by the fact that the property is restricted to non-nucleate microbes: the bacteria. Some of the complications of the genetics of higher plants and animals are thus avoided; such features as division of genetic information among many chromosomes, duplication of genetic information, and sexual exchange of genomes, rarely present problems with bacteria. A typical bacterial chromosome is a circle of DNA consisting of two intertwined circular molecules between which the basic genetic information of the organism is divided. Many bacteria carry one (or sometimes more) additional DNA circles called plasmids; these are much smaller than the chromosome, between 5 and 0.1% of its size and they supply supplementary information conferring properties such as resistance to drugs, toxin production, ability to transfer DNA to other bacteria or ability to utilize certain substrates (see section 6.1.2).

The techniques for studying microbial genetics are relatively simple, but an account would not be appropriate here and for detailed information the reader should consult a textbook of microbial genetics. It is worth recalling, however, that genetic information is encoded in fragments of the DNA called genes, the code being based on the detailed chemical structure of that particular portion of the DNA chain. The chemical structure is essentially a chain of components called bases, of which there are four types, trios of which specify which amino acids shall compose the protein coded for by a given gene. One gene generally codes for one peptide molecule (peptides are the chains of amino acids of which proteins are composed; many proteins consist of only one peptide). A change in the chemical structure anywhere along the DNA constituting a certain gene will cause mis-reading: either the wrong peptide will be formed or none at all. Such a change is called a mutation. Mutations occur spontaneously in about one in 10 million to 100 million of a normal bacterial population; their frequency can be increased considerably by treatment of the microbes with certain chemicals or with ionizing radiation. The study of mutants is basic to microbial genetics; not only are mutations invaluable markers for identifying genetic types of microbe, they can also be used to obtain maps of chromosomes, to study the regulation of genes (i.e. the control of the extent to which genetic information is used by the cell) and even to generate organisms with entirely new properties. In this chapter the application of modern methods of microbial genetics to the study of nitrogen fixation will be discussed briefly.

6.1   The nitrogen fixation gene cluster

The set of genes carrying the genetic information which enables bacteria to fix nitrogen is given the shorthand name *nif* by microbial geneticists. At this stage it will be useful to make a piece of conventional nomenclature clear. For convenience of discussion, all microbial genes have shorthand, three-letter names of this kind. Any mutation in the gene which leads to failure of its expression is given a minus sign: a mutant with mutated *nif*, which can no longer fix nitrogen, is called *nif⁻*. However, mutations in certain other genes can cause failure to fix nitrogen so, if a mutant is obtained which cannot fix nitrogen but which has not been positively identified as mutated in *nif*, it is termed Nif⁻ (a capital letter and no italics). Correspondingly, if a mutant which is *nif⁻* is treated genetically to restore its nitrogen-fixing capacity, yet there is uncertainty whether the original *nif⁻* mutation has been reversed, it is called Nif⁺. In more technical terms, *nif⁺* and *nif⁻* refer to wild type and mutant genes specifically; Nif⁺ and Nif⁻ refer to ability or otherwise of mutants to fix nitrogen, whether or not mutations in the *nif* genes are responsible.

Though Nif⁻ mutants of *Azotobacter vinelandii* and *Clostridium pasteurianum* had been reported in the 1960s, study of the genetics of *nif* really got started with some gene transfer experiments early in the 1970s. *nif⁻* mutants of *Klebsiella pneumoniae* were obtained and, by conventional methods of gene transfer, *nif* from wild type *K. pneumoniae* was introduced into the *nif⁻* mutants and their mutations corrected. From being Nif⁻ they became Nif⁺. The two procedures used to make those gene transfers are *transduction* and *conjugation*.

6.1.1   Transduction

Bacterial viruses are called bacteriophages. Certain of them, when they infect a new host, can transfer some of the genetic material of the old host with them. The co-transfer process is called transduction. If the new host survives the virus infection, and a few cells always do, it now possesses some genetical material from the virus's previous host. To transfer *nif* to *nif⁻* recipients, a *nif⁺* *K. pneumoniae* was infected with a bacteriophage called P_1. The virus was recovered from the infected host and used to infect *nif⁻* mutants. A proportion of these became Nif⁺, indicating that sometimes *nif* genes from the *nif⁺* type had co-transferred with P_1.

6.1.2   Conjugation

Certain plasmids confer upon their host the ability to donate the plasmid to a new host by a sort of pseudo-sexual process called conjugation. The donor cell carrying the plasmid (the male) actually comes alongside the recipient (female) organism, a tube grows between them and a replica of the plasmid forms and is transferred to the recipient. This, therefore, becomes a male and can donate the plasmid to yet another recipient. Not all plasmids, by any means, are self-transmissible (Tra⁺), but those that are can sometimes co-transfer other plasmids, or fragments of the donor's chromosomal DNA, to the recipient, as the bacteriophage did in transduction. To transfer *nif* to *nif⁻* recipients, a plasmid carrying both *tra* and resistance to the antibiotic kanamycin was introduced from its natural host, *Escherichia coli*, into wild type *Klebsiella pneumoniae*. The *K. pneumoniae* then became 'male'. It was 'mated' with *nif⁻* mutants of *Klebsiella* and the progeny acquired resistance to kanamycin, as would be expected when the mutants received a replica of the plasmid. A small proportion of these had also become Nif⁺, showing that the plasmid had, at low frequency, mobilized *nif* DNA from its host's chromosome and co-transferred it to the mutants.

6.1.3   Transformation

A third gene transfer process has more recently been used in the study of *nif*. By taking bacteria such as *E. coli* or *Bacillus megaterium* of a suitable

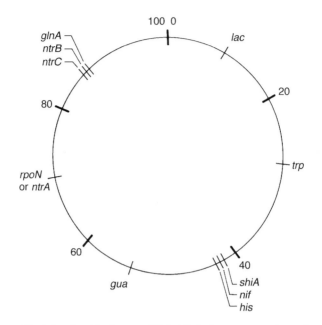

Fig. 6.1. A stylized map of *Klebsiella*'s chromosome shows the relative position of the nitrogen fixation genes. Several of the other genes quoted are positioned by analogy to the closely related *Escherichia coli* because their location in *Klebsiella* is not certain. The numbers are relative distances out of 100 units from the point at which replication is initiated (0). *lac*, Genes for a lactose-utilizing enzyme; *trp*, genes for synthesis of tryptophan; *shiA*, a gene concerned in uptake of shikimic acid; *nif*, genes for nitrogen fixation; *his*, genes for synthesis of histidine; *gua*, genes for synthesis of guanine; *rpoN* (*ntrA*), *ntrB*, *ntrC*, genes concerned in nitrogen regulation (see text); *glnA*, a gene for synthesis of glutamine synthetase.

age it is possible to 'force' them to take up raw DNA (plasmid or chromosomal) prepared in the laboratory. Recombinant DNA plasmids, made in the laboratory (section 6.4), are generally introduced into their hosts in this way.

6.2   The genetic map of *nif* in *Klebsiella pneumoniae*

Gene transfer experiments can give information about the location of genes on the chromosomes. For example, if the *nif*⁻ recipients of *nif* also had a mutation in a gene cluster concerned with making the amino acid histidine (a gene cluster called *his*), *his⁻* and *nif⁻* mutations were corrected simultaneously fairly often in gene transfer experiments. But if the *nif⁻* recipient also had a mutation (*trp*) in a gene cluster concerned with making the amino acid trytophan, the simultaneous correction of *trp⁻* and *nif⁻* was an extremely rare event. It follows that *nif* lies fairly close to *his* on the *Klebsiella* chromosome and a long way from *trp*. Numerous experiments of this kind converged to give the coarse 'map' in Fig. 6.1; the genes closest to *nif* are *his* on one side and *shi* on the other.

It is also possible to use gene transfer experiments to obtain a finer map. Researchers soon realized that the *nif* cluster must consist of several genes: at least two for the different polypeptides of the Mo–Fe-protein and one for the Fe protein (see Table 2.1) as well as at least one to regulate formation of enzyme according to whether ammonia is present or not (see section 3.6). Mutants in *nif* are now known which lack the Mo–Fe-protein, the Fe protein, or are *nif⁻* for other reasons. Complementation studies with *nif* plasmids (see section 6.4) indicated that mutations can occur in at least 15 distinct genes. By transducing these *nif⁻* mutations from one to another, always in a klebsiella with a *his⁻* mutation, one can arrange them in order relative to the *his⁻* mutation and to each other. Such transductional mapping contributed the bulk of the map of *nif* given in Fig. 6.2, in which the capital letters represent the individual genes which make up the cluster, but two genes, X and Y, were discovered as a result of cloning (see section 6.6) individual genes from the *nif* cluster and the existence of three, Z, W, and T, remained unsuspected until the complete DNA sequence of *nif* became available.

An important feature of the map in Fig. 6.2 is the division of the *nif* cluster into subclusters or operons. These are groups of genes, or single genes, which are 'read' to some extent independently of each other. The YTKDH operon carries the genes specifying the enzyme proteins themselves and there is now evidence that they are 'read' by the cell's genetic apparatus in the direction $H \rightarrow Y$ (how this sort of information was discovered will be indicated in the next section). The operons constituting *nif* are not all read in the same direction.

All the products coded for by all the *nif* genes have been identified and the amino acid sequences of all the proteins are known. By studying the behaviour of strains carrying mutations in various genes the functions of

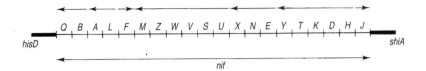

Fig. 6.2.   Finer map of the nitrogen fixation genes of *Klebsiella pneumoniae*. The capital letters indicate the separate genes that make up the *nif* cluster; for their functions see the text. The upper arrows indicate the subclusters (operons) in which genes are sequentially transcribed as a group, and the directions in which they are read.

several are well established and there are hints of what most of the rest do. They are summarized in order below, starting with the gene on the right hand of Fig. 6.2.

nifJ codes for an enzyme (pyruvate oxido-reductase) which generates from pyruvate the electrons used for the reduction of dinitrogen.

nifH codes for a peptide molecule, two of which combine to form dinitrogenase reductase, the smaller of the proteins constituting nitrogenase.

nifD codes for a peptide molecule (called the α-subunit), two of which combine to form part of dinitrogenase, the molybdoprotein of nitrogenase.

nifK codes for a somewhat different peptide molecule (called the β-subunit). Two of these combine to form the other part of the dinitrogenase protein, which is thus an $\alpha_2\beta_2$ tetramer composed of two pairs of similar peptides.

nifT codes for a very small protein molecule of unknown function.

nifY codes for a protein of uncertain function.

nifE codes for a peptide molecule, two of which combine to form the NifE product.

nifN codes for a peptide molecule, two for which combine to form the NifN product. The NifE and NifN proteins combine together to form a protein somewhat resembling dinitrogenase which appears to act as a template for synthesizing FeMoco.

nifX codes for a protein which may have a regulatory function.

nifU codes for a protein possibly concerned with the iron–sulphur centre of dinitrogenase reductase.

nifS codes for a protein which may also be concerned with the Fe–S centre of dinitrogenase reductase.

nifV codes for a peptide, two molecules of which combine to form an enzyme which makes homocitrate, a part of FeMoco, from 2-oxo-glutarate (section 1.3) and acetyl-coenzyme A.

nifW codes for a protein necessary for full activity of dinitrogenase.

nifZ codes for a protein which seems to be involved in the insertion of FeMoco into dinitrogenase.

nifM codes for a protein concerned with rendering dinitrogenase reductase active.

nifF codes for the flavodoxin which accepts electrons from pyruvate, *via* the *nifJ* product, and passes them to dinitrogenase reductase.

nifL codes for a regulator protein which switches off the whole *nif* cluster.

nifA codes for a regulator protein which switches on the whole *nif* cluster.

nifB codes for a protein involved in the synthesis of FeMoco.

nifQ codes for a protein involved in the uptake of molybdenum for FeMoco synthesis.

6.3 Genetic regulation of *nif*

The two genes *nifA* and *L* (Fig. 6.2) control whether nitrogenase is synthesized or not. Expressed more formally, they regulate 'reading' of the other operons. A mutation in *nifA* prevents synthesis of both nitrogenase proteins as well as all known products of other *nif* genes. One can introduce normal *nifA* into such a mutant on a plasmid (section 6.4) and reading of *nif* is restored; it follows that *nifA* makes a product which 'switches on' reading of the other operons. What this means is that *nifA* codes for a protein which is able to react with a part of the DNA constituting each operon and to cause the information in that operon to be made use of. The site of action of the *nifA*, product, called by geneticists an activator, is known as the 'promoter region' of the DNA; it has a special structure which NifA, the protein coded for by *nifA*, reacts with and distorts, enabling the genetic reading machinery to bind to the operon and start transcribing its genetic code. Rather more subtle experiments have shown that the product of *nifL* 'switches off' reading of *nif*, so *nifA* and *nifL* between them provide fine control of the extent to which *nif* is read (illustrated in Fig. 3.5).

Availability of ammonia is one external parameter which determines *nif* expression; how does its concentration affect the relative levels of the *nifA* and *nifL* products? The machinery is elaborate and not yet wholly understood, but much is nevertheless known. Classes of Nif⁻ mutants have long been recognized which map outside *nif*; one group lies close to *glnA* (see Fig. 6.1). They are called *ntr* (= nitrogen regulation) genes and *ntrBC* is actually contiguous with *glnA* in a cluster *glnAntrBC*. A gene earlier known at *ntrA* but now renamed *rpoN* lies on its own. In outline, the product of *rpoN* is essential for the action of the *nifA* product: in the absence of RpoN, NifA cannot switch on *nif*. In addition, the product of *ntrC* is also necessary for the *nifLA* operon to be read (i.e. for NifA and NifL to be made at all). Thus, both *rpoN* and *ntrC* must be working for *nifLA* to do its job. The formation of the *ntrC* product (but not of the *rpoN* product) is regulated by ammonia: NtrC is not formed if plenty of ammonia is present. In a manner that recalls the action of NifA mentioned in the preceding paragraph, NtrC binds to the promoter region of *nifLA* DNA, distorting it and enabling it to be read and transcribed. Recalling the way *nifL* functions, the product of *ntrB* interferes with NtrC's action. It thus seems that the *ntr* genes provide a coarse control of *nif*, switching it on or off, and the *nifLA* system provides the fine tuning. A fascinating feature is that the *ntr* system not only controls *nif* but also a variety of other aspects of the nitrogen metabolism of *Klebsiella* including its ability to obtain N from urea, amino acids and nitrate. The *rpoN* product proves to be a component of the apparatus for transcribing many genes and operons concerned with N-metabolism: it is a so-called σ-factor.

Of course, the *ntr* story has only pushed the original question back a stage, because one can still ask how does the *ntr* system respond to ammonia? Much of the answer is known, but discussion of the matter leads into the subject of nitrogen regulation in general, which is outside the theme of this book. For present purposes the important conclusion is that *nif* genes, when present, come under a coarse control system common to several aspects of N metabolism. It follows that it is possible to mutate the *ntr* system and obtain strains which escape ammonia repression: they are called 'constitutive' Nif$^+$ mutants because they make nitrogenase even if ammonia is present.

As indicated in section 3.1.2, oxygen is also a regulator of *nif* expression. In this case the regulatory process is somewhat more direct in that it bypasses the *ntr* system. Once again the product coded for by the *nifL* gene is the key regulator: mutants defective in *nifL* can actually synthesize nitrogenase in the presence of oxygen, although it is then useless. The NifL protein proves to be an oxido-reducible protein, of the class called flavoproteins, and is thus capable of responding to oxygen levels. In the reduced state, when the oxygen concentration is low or zero, it does not interfere with *nif* expression, but its oxidized form prevents the NifA protein from distorting promoter DNA and allowing translation.

In section 3.3 the physiological consequences of nitrogenase's need for ATP were discussed. That need has genetical repercussions, too, for it would obviously be disadvantageous for the cell to divert valuable ATP to nitrogen fixation when food is short. So the expression of *nif* is regulated in response to the cell's energy supply. The NifL protein is also involved here. Energy-starved cells have less ATP and more ADP than well-provided cells; ADP reacts with the reduced form at NifL and causes it to block promoter activation by NifA.

6.4   *nif* plasmids

Section 6.1 described how a plasmid was used to mobilize the *nif* genes from the *Klebsiella* chromosome and transfer them to *nif*$^-$ mutants. The closeness of *nif* genes to *his* was also emphasized. Since the original plasmid came from *E. coli*, it would readily transfer back to its original host and one might hope that, albeit at low frequency, it might co-transfer *nif* genes into the organism. If the *E. coli* could use the *nif* genes, an entirely new species of nitrogen-fixing bacteria would thus have been generated. My colleague, R. A. Dixon, attempted this experiment early in the 1970s. In outline, a specially chosen *E. coli* strain with a *his*$^-$ mutation was mated with the 'male' nitrogen-fixing *Klebsiella* described in section 6.1 and His$^+$ progeny were obtained at a low frequency. Nearly all of these, to our delight, were able to fix nitrogen: the *nif* genes had co-transferred with *his* and the *E. coli* was able to use them. Ammonia

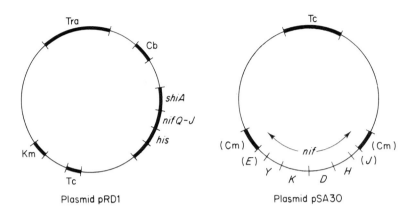

Fig. 6.3. Speculative maps of two plasmids carrying nitrogen fixation genes. The relative positions of the genes are not established in pRD1 except for those including *his* and *nif* from *Klebsiella*. Abbreviations are as for Figs. 6.1 and 6.2 except: Tra, genes for self-transmissibility; Cb, gene conferring resistance to carbenicillin; Tc, gene conferring resistance to tetracycline; Km, gene conferring resistance to kanamycin; Cm, gene conferring resistance to chloramphenicol. Plasmid pSA30 has some *nif* genes spliced into the Cm gene of a plasmid called pACYC184. Incomplete genes are in parentheses.

repressed *nif*, so the *E. coli* must have had the right sort of *ntr* system to 'switch on' *nif*.

A close examination of the nitrogen-fixing *E. coli* hybrids showed that two classes of event had occurred. (1) A strain had arisen in which *his* and *nif* from *Klebsiella* had integrated completely into the *E. coli* chromosome. It was genetically a new type of nitrogen-fixing organism. (2) Two strains had arisen in which the *Klebsiella* DNA had not integrated into the chromosome, but had formed new plasmids; not just one, but two in one case and three in another. The *his* and *nif* genes were on one of these.

The spontaneous formation of plasmids carrying *nif* in *E. coli* lent credibility to an idea for genetic manipulation of *nif*: could one not build *nif* into a self-transmissible plasmid? Dr Dixon successfully constructed such *nif* plasmids; a sketch map of a particularly useful one pRD1 is in Fig. 6.3. The plasmid was constructed in *E. coli* and carries three drug resistance genes which are useful as genetic markers in the laboratory.

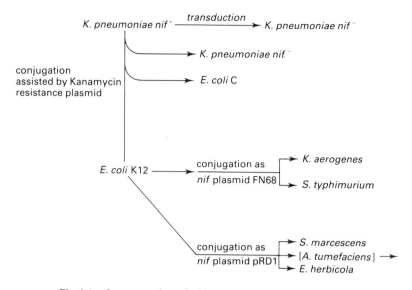

Fig. 6.4.   Some transfers of *Klebsiella pneumoniae* nitrogen fixation genes to new bacterial hosts. For details see sections 6.1.1, 6.1.2 and 6.4.

Plasmid pRD1 is so stable in that 'host' that one can make mutations in its *nif* cluster. One can then transfer the mutated plasmids back to *nif⁻* mutants of *Klebsiella* and judge whether the mutations are the same or not. This is an example of a genetical test called 'complementation' and was the way in which 15 of the *nif* genes were first identified. Plasmid pRD1 has been used for a variety of research purposes including the construction of new types of diazotrophs (section 6.5 below); it was also the source of the primary *nif* gene clones discussed in section 6.6.

6.5   New nitrogen-fixing bacteria

The nitrogen-fixing *E. coli* hybrids described in the previous section were the first types of new nitrogen-fixing microbes to be developed deliberately by genetic manipulation. This success opened up some of the future possibilities discussed in Chapter 7. Numerous transfers of *K. pneumoniae nif* have now been made using such plasmids, some of which are illustrated in Fig. 6.4.

Bacteria such as *E. coli*, *Salmonella typhimurium*, *Serratia marcescens*, *Erwinia herbicola* or *Klebsiella aerogenes* become perfectly effective nitrogen-fixing bacteria when provided with *nif*, though they still only fix anaerobically. Equally interesting are organisms such as *Agrobacterium tumefaciens*, which does not fix nitrogen even if it carries a plasmid bearing *Klebsiella*'s *nif* genes. It can be shown that the agrobacterium recipients of the plasmid have not excised *nif*, because they will transfer the plasmid to *E. coli*, which can then fix nitrogen quite readily. *A. tumefaciens* lacks some supplementary information which is necessary for complete expression of *nif*. Another interesting case is that of *Proteus mirabilis* which does not express *nif* at all with pRD1 (actually a plasmid derived from pRD1) but does so if it is provided with an additional constitutive *nifA* gene: it seems to lack part of the *ntr* system.

Reports have appeared in the scientific literature of the transfer of *nif* genes from *Rhizobium leguminosarum bv. trifolii* to a strain of *Klebsiella aerogenes* which does not normally fix nitrogen, of the transfer of *nif* genes between photosynthetic bacteria, and of the transfer of *nif* genes from *B. japonicum* to *Azotobacter vinelandii*.

6.6   Artificial *nif* plasmids and the physical map of *nif*

In the mid 1970s, techniques were developed for genetic manipulation in the test tube: DNA could be isolated from a source, cut into pieces (figuratively speaking) with special enzymes (restriction enzymes) and reassembled with other enzymes (ligases). If DNA from different organisms were cut, mixed and reassembled, all sorts of mixed genetic elements could be constructed artificially. If, finally, part of the DNA came from an identifiable genetic unit such as a bacteriophage (see section 6.1.1) or a plasmid, one could construct bacteriophages or plasmids carrying alien genetic information and these would multiply in a host such as *Escherichia coli*, producing numerous copies of themselves and therefore of the new DNA. Under the name of 'genetic engineering' (a name which had been used before for more conventional genetic manipulations) this technology achieved brief notoriety because, by careless insertion of alien genes (e.g. genes specifying the ability to cause disease or cancer) into the genetic complement of *E. coli*, one might make it into a really nasty microbe. Yet one could equally introduce highly desirable properties, such as ability to make rare hormones, into bacteria, which could then be grown in large quantities. The *nif* gene cluster of *Klebsiella pneumoniae* has proved amenable to such artificial manipulation but it is cut itself by most restriction enzymes used so far. Nevertheless, all the *nif* genes have now been spliced into small plasmids, usually in groups, sometimes as single genes, some including part of *his*, too. Often these plasmids form many copies of themselves per bacterium. Thus great quantities of *nif* DNA in a relatively purified form have become available for research; the DNA in such plasmids is said to be 'cloned'.

Artificial plasmids made in this way are called 'recombinant' plasmids by molecular biologists; Fig. 6.3 includes a sketch of a particularly useful one, pSA30, which carries the genes coding for nitrogenase peptides together with their promoter region. Comparable plasmids carrying all the *nif* operons have been made, as well as plasmids carrying just the promoter regions, others without their promoter, yet others carrying only part of a gene. Recombinant plasmids have also been made in which entirely alien genes such as *lac* (see section 6.6.3) have been fused into the *nif* genes, or attached to *nif* promoters, and others in which *nif* genes have been put under control of non-*nif* promoters. There are many ways in which the recombinant DNA technology can be exploited; five examples are listed below.

6.6.1   Mapping

When DNA is cut with a restriction enzyme the fragments are of different sizes and can be distinguished by electrophoresis. Now, there exists a class of mutation (an 'insertion' mutation) in which an extra piece of DNA has become inserted into a *nif* gene so, when that DNA is cut by a restriction enzyme, the fragment comes out larger than it would otherwise be. Analysis of a variety of insertion mutants gave a physical map of *nif* matching that in Fig 6.2. Cloned *nif* genes can also be made to express their products independently, and those products can be characterized; it was during that kind of research that the two extra genes, X and Y, were discovered.

6.6.2   Constitutive constructs

The *nif* promoters are regulated by ammonia through the *ntr* system (see section 6.3). By attaching *nifA* to a different promoter, namely one controlling a drug resistance, a plasmid was made which made *nifA* product whether ammonia was present or not. Such a plasmid was used to cause *Proteus mirabilis* carrying a *nif* plasmid to express *nif* (section 6.5).

6.6.3   Fusions

The *lac* gene of *E. coli* codes for an enzyme, β-galactosidase, which hydrolyses lactose. It also hydrolyses certain analogues of lactose to give products which are coloured. Clones of *lac* exist, as well as a bacteriophage carrying *lac*. Such clones can be manipulated so that the *lac* gene becomes fused to a *nif* promoter and formation of the colour controlled by it. In this way the roles of *nifA* and *nifL* in controlling the other operons of *nif* was worked out, and also the effects of *ntr* gene products on *nifA* and *L*.

6.6.4   DNA sequences

Methods exist to determine the actual sequences of bases in the DNA of genes. Cloned *nif* genes have been sequenced and, by using the 'genetic code' to read the information contained therein, the amino acid sequences of their products have been deduced. Thus the precise sequence of a protein is sometimes known without its properties or function being at all clear!

The *nifW* and *nifZ* genes were revealed by sequencing, and in 1988 researchers in Germany published the DNA sequence of the whole *nif* cluster and discovered its remaining gene, *nifT*. DNA sequencing has also been invaluable for elucidating the inter-genic regions which are involved in regulating gene expression, in particular for identifying the places on the DNA chain where regulator substances bind and the ways in which they influence its folding.

6.6.5   Probing

DNA consists of two intertwined chains of bases. If purified DNA is heated to 70 – 80 °C, the chains unwind, but they reassemble on cooling. If the unwound, single-strand, DNA is mixed with other DNA with which it has base sequences in common, the homologous sequences will stick together. Make one of the DNA strands radioactive (which can be done in a variety of ways) and you have a 'probe' which will stick to heated homologous DNA and make the pair radioactive when cooled. Radioactive probes from pSA30 have shown that the nitrogenase DNA from all sorts of diazotrophs is very similar: cyanobacteria, rhizobia, azotobacters and so on. The *nifH* gene is a particularly good probe of this kind. It is obvious that such a probe enables one to 'spot' *nif* genes in other organisms and, in this way, the genetics of *nif* in a number of bacteria have been opened up.

6.7   Genetics of *nif* in other diazotrophs

Recombinant DNA technology has provided a means of studying the genetics of almost any diazotroph, although regular *nif* mutants of many such organisms have also been very useful. In *Azotobacter vinelandii*, for example, well-characterized *nif* mutants have long been available, some lacking the Fe-protein, some lacking the Mo–Fe-protein, some lacking both (presumably mutated in the regulatory apparatus) and at least one constitutive mutant which expressed *nif* in the presence of ammonia. But progress with the detailed genetics of *nif* in *A. vinelandii*, and its 'cousin' *A. chroococcum* in which *nif* mutants were also available, was slow because of the absence of reliable quantitative procedures for genetic analysis, such as transduction and conjugation. The advent of *nif* probes from *Klebsiella pneumoniae*, and means of cloning *Azotobacter* DNA, dramatically altered the position. Most, probably all, of *nif* has been cloned into *E. coli* from *Azotobacter chroococcum*: it seems to have a partial *nif* cluster in its chromosome which is rather like that of *K. pneumoniae* with the component genes in a similar order but more widely separated. But some of its *nif* genes are outside the main cluster altogether. It also has a second copy of *nifH*, which is also remote from the main *nif* cluster. This is one of the *vnf* genes which codes for the alternative nitrogenase in *A. chroococcum* (see section 2.4, 3.5 and 6.8).

Azotobacters seem also to have the equivalent of *Klebsiella*'s *nifF*, but a *nif* gene resembling *Klebsiella*'s activator gene *nifA* seems to be remote from the main cluster. The *nif* genes of azotobacters are, like those of almost all other diazotrophs, tightly regulated by ammonia and, not unexpectedly, a gene homologous to *ntrA* has been found in *A. vinelandii*. One homologous to *ntrC* is also present but, surprisingly, it differs from *ntrC* of *K. pneumoniae* in that it does not regulate *nif*, so the *nif* regulatory apparatus of azotobacters is different. *Azotobacter* is an especially interesting genus, not only because of its alternative nitrogenase but also because of its efficient physiological adjustment to oxygen. The genetic basis of oxygen exclusion has begun to be disclosed with the discovery and study of oxygen-sensitive (Fos⁻) mutants of *A. chroococcum*.

Little can be said at present about the genetics of the aberrant nitrogenase of *Streptomyces thermoautotrophicus*: DNA probing has revealed no genes homologous to *nifH* or *nifKD* of *K. pneumoniae* in its genome.

In a much-studied strain of the photosynthetic cyanobacterial diazotroph *Anabena* called PCC7120, the vegetative cells carry the nitrogenase genes but do not express them. It transpires that *nifK* is separated by a length of DNA from *nifDH*, which are contiguous. As the heterocysts of this strain develop (see section 3.1.7), the intervening DNA becomes excised so that a *nifKDH* operon like that of *Klebsiella* is formed. The *nif* is duly expressed.

A different heterocystous species, *Anabena variabilis*, whose filaments are able to fix nitrogen if the oxygen tension is very low, has two sets of *nif* genes, one expressed in the heterocysts when oxygen is abundant, the other expressed in the filaments when oxygen is sparse. Probing for nitrogenase genes has revealed genes homologous to *Klebsiella*'s *nifK, D, H* and other nitrogen fixation genes in all diazotrophs so far examined, including non-cyanobacterial photosynthetic bacteria, the sulphate-reducing bacteria, the thiobacilli, *Frankia*, and even the methanogenic bacteria, which are Archaea (see section 4.1). Some, such as certain rhizobia and photosynthetic bacteria, seem to have more than one copy of their presumptive *nif* genes; *Clostridium pasteurianum* has a surprising six homologues of *nifH*.

Of special interest are the rhizobia, whose nitrogen fixation genes display a wide variety of arrangements and which, naturally, need additional genetic information to equip them for entering into the legume symbiosis. As far as their *nif* genes are concerned, *Rhizobium leguminosarum* and its *trifolii* variety have single copies of *nifH, D* and *K* but its *phaseoli* variety has two copies of *nifK* and *D* and three of *nifH*. An unexpected discovery made at the end of the 1970s using the *nif* plasmid pSA30 as a probe was that, unlike *K. pneumoniae*, all three varieties carry their *nif* genes on plasmids. In contrast, the *nif* genes of bradyrhizobia are generally chromosomal. In none of the rhizobia do the *nif* genes form the tidy cluster found in *K. pneumoniae*; they are dispersed in groups among the rest of the microbe's genome. Sometimes they are closely associated with genes called *sym* (for symbiosis) which determine host specificity, or with *nod* genes (for nodulation) or *fix* genes (for nitrogen fixation); *fix* genes may be as yet unrecognized analogues of *nif* genes. These, too, are often on plasmids, sometimes but not necessarily the same ones that carry *nif* genes. It is possible to cause clover rhizobia to nodulate peas by genetically manipulating into them the *sym* plasmid of the pea rhizobia – but the nodules so formed prove to be ineffective.

The establishment of a legume symbiosis involves the co-ordinated expression of perhaps 50 genes. Some, such as those coding for leghaemoglobin (see section 5.1), are carried by the plant, others by the bacteria. Their gene products are called nodulins, and these substances, released in controlled order, exert sequential influences on the infection of the plant, on the migration of the bacteria, on the development and number of nodules, and on the changes in physiological status whereby the rhizobia become bacteroids. For example, a crucial gene is rhizobial *nodD*. This codes for a protein which recognizes the flavonoid secreted by the host plant and then prompts the bacteria to migrate appropriately and to set about colonization. The flavonoid actually participates in switching on other rhizobial genes, including one which codes for a lipochitooligosaccharide molecule which prompts the plant to initiate nodule growth. A finely adjusted give-and-take takes place between plant and bacterial genomes which leads to profound changes in the physiologies of both partners. Understanding in this area has advanced impressively over the past decade and recent specialist literature should be consulted for details.

6.8   The *vnf and anf* genes

It was genetical research that, in the early 1980s, settled the then controversial question of the existence of alternative nitrogenases (see section

2.5). These enzymes are specified by structural genes which resemble *nif* closely enough to be recognized by *nif* probes such as pSA30 but which, for clarity, have been given new shorthand names. Genes which code for vanadium nitrogenases are termed *vnf* genes, and *anf* is used to refer to genes which code for the third, iron-based nitrogenase. The letters used with these names correspond as far as possible to those used with *nif*: thus *vnfD, K* and *H* signify the structural genes of vanadium nitrogenases. However, in *Azotobacter chroococcum* a small gene, *vnfG*, lies between *vnfD* and *vnfK* (it codes for the low molecular weight ferredoxin-like δ-subunit of vanadium nitrogenase) and *vnfH*, although located close to *vnfDGK*, is not contiguous with them as *nifH* is in *Klebsiella*. The third nitrogenase does have contiguous structural genes, arranged as *anfHDGK*. Both the *vnf* and *anf* systems have activator genes *vnfA* and *anfA*, corresponding functionally to *nifA*, but they share certain regular *nif* genes such as *B, N* and *V*; in addition, *anf* shares with the *vnf* system a pair of linked genes, *vnfEN*. The genetics and regulation of the alternative nitrogenase systems are thus a little complicated and details have not yet been elucidated completely, but both similarities to and divergences from regular *nif* are already evident.

6.9   Evolution of nitrogen fixation

It is worth stepping back from genetics to consider the implications of recent biochemical and genetic findings on our understanding of the origin and evolution of diazotrophy.

As far as the symbioses are concerned, the nodular diazotrophic symbioses among the flowering plants or angiosperms fall into two large groups: the legumes with their rhizobia and the actinorhizal trees and shrubs with their frankias. For most of the twentieth century botanists regarded the two groups as wholly unrelated to each other, but in the 1990s molecular genetical studies revealed that they are more closely related than was hitherto thought. Comparisons of DNA sequences within the plant genomes indicate that the two groups form a phylogenetic clade; that is, they have a common evolutionary ancestor not shared by other angiosperms. Thus, it is probable that the ability of angiosperms to enter into those symbioses emerged but once in evolutionary time. Naturally particular symbioses could subsequently have appeared – or disappeared – at any time, and would have shown normal evolutionary divergencies.

The nodular symbioses most likely arose from casual associations of plants with certain free-living diazotrophic bacteria which were ancestral to their present day endophytes. In the case of the rhizobia, a most interesting discovery has been the fact that the *sym* and associated genes are on plasmids. These genetic elements could be transferred between species far more readily than could chromosomal genes during evolutionary time. The chromosomal background of *Rhizobium* species is closely related to that of the genus *Agrobacterium*, species of which cause crown gall growths on susceptible plants. And the genetic determinants for crown gall pathogenicity are also plasmid-borne. So it is tempting to regard the two bacterial genera as respectively mutualistic and pathogenic descendants of a common plant-associated (commensal or pathogenic) ancestor. The legumes originated rather early in the evolutionary history of the flowering plants, around 200 million years ago. Nodules are too soft to have left recognizable traces in the fossil record, but it is fair to assume that rhizobial symbioses emerged later than this. Some authorities consider that uneven distribution of nodulation among the three subfamilies of legumes (it is found in some 85% of Papilionoidae, 23% of Mimosoidae and few Caesalpinoidae) implies that rhizobia originated after these groups had diverged, less than 100 million years ago. Such ages are relatively recent in terms of the billions of years of bacterial evolution.

In contrast, the cyanobacterial symbioses must surely have originated independently of the nodular symbioses, and probably much earlier because their plant hosts span the whole range of complexity, from primitive to sophisticated. They too could reasonably have arisen from casual mutualistic associations of plants with diazotrophic cyanobacteria; the curious fact that the cyanobacterial endophytes in the coralloid roots of *Macrozamia* retain chlorophyll while rarely if ever encountering light would then be simply another example of a vestigial property which has not yet been lost.

As far as the nitrogen-fixing enzyme system itself is concerned, the existence of three classes of nitrogenase, those coded for by *nif, vnf* or *anf* genes, presents another evolutionary problem. Which, if any, came first? Though the molydbenum nitrogenase appears to be the most widespread today, as well as being genetically dominant and biochemically the most efficient, one must remember that, since the mid 1930s, when the importance of molybdenum was first appreciated, microbiologists have almost always added molybdate to their culture media, thus ensuring that their isolates had been subject to selection against diazotrophy based on the other two nitrogenases. The distribution of the nitrogenases among laboratory strains may not accurately reflect their distribution in nature; the question is still open at the time of writing.

The ways in which bacteria, especially aerobes such as *Azotobacter* and heterocystous cyanobacteria, developed means of utilizing the highly oxygen-sensitive nitrogenase in the presence of air presents yet another fascinating topic in physiological evolution. However, perhaps the most fundamental evolutionary question concerns the origin of diazotrophy itself. Although there is very little palaeological evidence to go on, some instructive speculations are possible. Thus, for many years scientists inclined to the view that nitrogen fixation was an ancient property, one which emerged soon after, if not along with, the origin of bacteria themselves, over 3 billion years ago. This opinion arose largely because diazotrophy is only found among the most primitive of living things – the bacteria – and even among those it seemed most prevalent among the more primitive types, the anaerobes. If it was an ancient property, it must be one which was readily lost during the development of higher organisms. Some support for this view comes from the detection of traces of what may have been cyanobacteria with heterocysts in very ancient pre-Cambrian rocks (over 3×10^9 years old). On the other hand, what good would heterocysts and nitrogenase have been in those days when, according to geochemical dogma, the planetary atmosphere and environment contained no oxygen but some free ammonia and oxidized nitrogen? Could the enzyme in fact have been doing something quite different, such as removing toxic cyanides? Yet if the enzyme is so old, why has there been so little evolutionary divergence in its structure? It is much the same wherever it comes from, and the nitrogenase genes seem to be much the same even in the evolutionarily distinctive Archaea. And why have no plants or fungi learned how to use it? After all, the properties of cyanobacteria or azotobacter tell us that there is no serious obstacle to nitrogen fixation by ordinary aerobic creatures. An alternative view is that it is not an ancient property. The ability to fix nitrogen may have emerged late in evolutionary history: sometime after oxygen became a permanent component of the planetary atmosphere and after plants had used up all superficial supplies of ammonia, nitrates and other sources of fixed N. If it emerged once, in a bacterium, then we know that it could spread quite easily among other bacteria – perhaps even aided by plasmids – and this would account for the scattered way in which *nif* is today distributed throughout the microbial world. It may be significant in this context that some *nif* genes of rhizobia and some other diazotrophs occur naturally on plasmids. A recent origin would also account for the essential sameness of its principal product, nitrogenase, no matter where it comes from. It would follow that the property is still spreading and that, even without mankind's interference, it might yet become a property shared by higher organisms. But if this view is true, what were those heterocyst-like bodies doing in pre-Cambrian rocks? Perhaps heterocysts were once not nitrogenase compartments but spore-like bodies? In this area, discussion can be endless, and the question of the age of the nitrogen fixation process is still open, but at least it is no longer wise to assume that nitrogen fixation was one of the pristine properties of living things.

7   The future

The importance of the nitrogen cycle in the productivity of the biosphere has caused many people, not only scientists, to focus their attention on nitrogen fixation, which is its rate-determining step (see section 1.1). In sections 1.7 and 1.8 some indication was given of the present situation regarding nitrogen fertilizer and world food supplies. To amplify some of the points made there, the human population of this planet has grown steadily over many millennia and has expanded dramatically during the twentieth century. Local episodes of famine, where the population of a locality exceeded the capacity of that locality to supply food, have occurred throughout history. Indeed, Thomas Malthus's now famous essays of 1798 and 1803 demonstrated the ultimate dependence of population on agriculture and gave the name 'malthusian' to this kind of anxiety. Historically, trade and technology combined to postpone, or at least to localize, the problems foreseen by Malthus until the mid twentieth century, when it became clear that the global population had reached such numbers that, setting aside wars and political problems, agronomists doubted the capacity of world agriculture to feed humanity as a whole. During the three decades following 1945 plant breeders came to the rescue: world food production kept ahead of a rapid population increase because high-yielding strains of rice, wheat and other cereals were bred and introduced, promoting sometimes spectacularly increased harvests in developing countries. But that so-called green revolution cannot be repeated and the world's population continuous to mount. In the year 2000 there will be about 6 billion people in the world, mostly young and fertile, and even if all birth control programmes are successful the number cannot but rise to some 8 billion by 2015–2020 – unless some unforeseen catastrophe intervenes. This figure is close to the 9 billion which is thought to be the most that modern 'hi-tech' agriculture can support, taking into account weather, water supplies and the total area of the fertile zones of this planet.

More detailed discussion of the complex human, environmental and ethical problems generated by the population explosion is inappropriate here, but no-one disputes that these people must be fed; therefore drastic changes in agricultural practice will be needed, with a high priority being assigned to maximizing nitrogen inputs into crops with minimum disturbance of water supplies and land quality. Considerable improvement can come from the application of existing knowledge such as that outlined in earlier chapters, but the probability is that this will provide no more than breathing space in which to develop longer-term solutions. Some of the more radical approaches which are possible are discussed briefly below.

7.1   Chemical prospects

In simplistic terms, an approximate doubling of the 1990 world-wide input of chemical produced N-fertilizers to some 1.6×10^8 tonnes N/year ought to solve the agronomic problem of N-supply, but such fertilizers already show a number of disadvantages from this point of view. They need a sophisticated industry, which implies capital costs out of reach of the most needy countries. Even though they are cheap at the factory gate their manufacture involves a significant cost in terms of global energy consumption and they also have high packaging and transport costs. They are wasteful in use, because only about 50% of the fertilizer applied to a soil is actually taken up by the crop, and that which escapes pollutes the environment in various ways. For example, eutrophication of inland and coastal waters can be caused by run-off of N-fertilizer; nitrate concentrations in drinking water are increasing everywhere and often exceed statutory limits even in developed countries; and microbial de-nitrification of the residual N-fertilizer leads to substantial release into the atmosphere of nitrous oxide, a powerful greenhouse gas. Nevertheless, chemical N-fertilizers work very well and there is no getting away from this fact. The high-yielding cereal crops which have underpinned the green revolution require high inputs of chemical fertilizers, especially nitrogen, and agronomists have calculated that well over a third of the world's present population is alive and fed by virtue of the Haber process. If N-fertilizers could be produced more cheaply, or produced by simple means appropriate to the poorest countries, and if their use could be managed so as to ensure more efficient uptake by plants, then the future of world agriculture would look somewhat brighter. Recent advances in understanding the basic chemistry of dinitrogen have opened several possibilities for modifying the traditional Haber process, or by replacing it by altogether simpler procedures, but none is yet close to practical exploitation.

7.2   Botanical prospects

So large an increase in global input of chemical N-fertilizer is not feasible either economically or environmentally, and it follows that biological nitrogen fixation will have to play an increasingly important part in future agricultural practice. As indicated in Chapter 1, the most useful nitrogen fixers in agriculture are the legumes. Beans, peas, pulses, groundnuts, either consumed as such or processed into prepared food, today contribute substantially to human diets all over the world. Clover, alfalfa (lucerne) and other legumes are used increasingly as fodder supplements and as green manures. Leguminous trees such as *Lucaena*, as well as trees such as alder (*Alnus*) which form actinorhizal symbioses, show promise in tropical and temperate forestry, respectively. *Azolla* (Fig. 7.1) is still very valuable in the cultivation of rice, and some of the more casual root associations discussed in Chapter 5 ought to become useful in cereal and cane sugar cultivation soon. Innovative exploitation of already known nitrogen-fixing systems such as these can be very effective locally, but in global terms their impact will probably be rather small. For the longer term, however, a few more radical prospects exist.

The scope for generating new plant hybrids was much widened in the 1970s by the exploitation of a non-sexual procedure which is called 'somatic cell hybridization'. Live plant cells, from a leaf for example, can be treated with enzymes which remove their cell walls without killing them; they can then be induced to fuse in pairs, and with appropriate plant hormones many of the fused products can be regenerated into whole plants. If cells from two different plants are mixed and fused, hybrids which would not be formed by sexual crossing can sometimes be obtained (hybrid petunias have been prepared in this way). Legumes do not, of course, hybridize with brassicas, but somatic hybridization might yet confer the property of nitrogen fixation on a large edible green

Fig. 7.1. *Azolla* is harvested from a rice paddy in Vietnam for use as a nitrogen-rich green manure. (Courtesy of Dr I. Watanabe.)

vegetable! More seriously, the prospect of using somatic hybridization to generate a cereal able to form effective rhizobial nodules like those of a soya bean has been taken seriously; the prospect is still remote, but cells of barley and soya bean have been induced to fuse.

Another approach is to induce cereals themselves to form effective nitrogen-fixing associations with rhizobia. Some rhizobia are fairly catholic in their host specificity, colonizing several species of legumes, and one type is known that colonizes a non-legume. Careful manipulation of plant cultivars, bacterial strains and infection conditions might yet generate a 'natural' symbiosis of this kind.

Finally, plant chloroplasts, which as organelles within the cell cytoplasm carry the photosynthetic machinery, are known to be the evolutionary descendants of endocellular symbiotic cyanobacteria. The possibility of generating crop plants, especially cereals, which carry within their cells comparable dependent diazotrophs, or organelle-like 'diazoplasts' derived therefrom, capable of regulated nitrogen fixation is an exciting if at present remote prospect.

7.3   Genetical prospects

All but the last of the prospects just mentioned leave the plant dependent on bacteria for its diazotrophy, and bacteria are notoriously unreliable, as the abundance of ineffective rhizobia in agriculture testifies. Modern molecular biological techniques would allow the genetic information for nitrogen fixation to be manipulated from a bacterium such as *Klebsiella*, where it exists as a convenient cluster of genes, and be transferred to a plant relatively easily. However, mere transfer of bacterial *nif* genes to the genome of a plant would not, in itself, give those plants the ability to fix nitrogen. (Indeed, given the intimacy of many nitrogen-fixing plant–microbe associations, it is probable that *nif* genes enter plant cells fairly often in nature when senescence or damage occurs.) The reason is basically that the plant's genetic apparatus cannot read the genetic information carried in bacterial DNA. Yet there are all sorts of ways in which the plant might be 'taught' to fix nitrogen. The *nif* genes could be introduced on a plant virus, together with supporting genes for regulating *nif*, for keeping oxygen out of the way and so on. Or the necessary genes for nitrogen fixation might be 'spliced' into benign parasitic bacteria which would persist inside or close to the plant. A more subtle approach would be to splice *nif* into one of the major plant organelles, chloroplasts or mitochondria, which retain some capacity to read bacterial DNA because they probably evolved from parasitic endophytic bacteria. Analogous diazotrophic organelles are perfectly conceivable – at least in principle.

One might even bypass plants and put *nif* genes into the sorts of bacteria which are symbiotic with animals: a nitrogen-fixing goat (its rumen populated with nitrogen-fixing bacteria) might flourish on a diet of paper even better than normal goats are reputed to do! Or one might use the microbes as little ammonia factories: make a constitutive Nif$^+$ strain (section 6.3) of a cyanobacterium, introduce another mutation so that it cannot use the ammonia it forms (such mutants have already been made in *Klebsiella*) and you have an organism that makes nitrogen into ammonia at the expense of solar energy.

7.4   Consequences

There are numerous bright ideas one may have for the future and they all have complications when one considers them in detail. But many are plausible in principle; the specialized literature should be consulted for amplification of some of those just mentioned. If, by some means or another, a successful breakthrough were achieved and (for example) a natural nitrogen-fixing cereal were developed, would it be wholly beneficial? Some environmentalists have expressed alarm: fears of rampant nitrogen-fixing weeds upsetting the world's ecological balance, for example. It is always wise to consider the possible risks of scientific innovation, and there are indeed risks in the indiscriminate spread of nitrogen fixation. But on consideration they prove to be minor ones, and the potential benefits are vast. Consider what would happen if nitrogen-fixing plants including cereals become a reality. Essentially, they could behave rather like soya beans or clover do today: nitrogenous matter would cease to limit their productivity and other minerals such as potassium, sulphur or phosphate would become limiting. But the manufacturing and transport costs of nitrogenous fertilizers would have been saved, as well as much of the environmental cost due to run-off of unused fertilizer into lakes and drinking water. If, by careful agricultural management, adequate K, S, P, etc., were supplied, then carbon dioxide would become the limiting nutrient: as with soya beans in glasshouse conditions today, photo-assimilation of CO_2 would limit productivity. (Intensive farmers would probably spend their money on enriching the atmosphere of huge plastic greenhouses with CO_2.) Clearly the prospect of a catastrophic environmental 'take-over' by nitrogen-fixing weeds and grasses is not to be taken seriously (it would probably have happened already if it were feasible). But in this life one does not get something for nothing: a nitrogen-fixing cereal would grow rather more slowly than its cousin being supplied with nitrate fertilizer. The reason is not because some of its ATP would be needed to make nitrogenase work (see section 2.2), for the reduction of nitrate has a similar energy cost, but rather because nitrogenase is a slow enzyme needed in large amounts, and the energy cost of making it and sustaining it could be quite a drain on the plant's economy. Looked at from another viewpoint, the northernmost limits of a nitrogen-fixing cereal would be a little further south than those of its cousins growing with nitrate. This price ought to be small compared with the savings in costs of artificial fertilizers.

There are also some minor hazards to be considered, though once recognized they could be easily guarded against. Several plant-pathogenic bacteria operate by rotting the living plant, a slow procedure because plants are generally not rich in fixed N. If such microbes picked up *nif* genes and could use them, they would become more pathogenic. Another hazard: ordinary lakes and rivers have very little fixed N. If one of the more obstinate water weeds acquired the ability to fix nitrogen, it might well clog up waterways and reservoirs seriously, just as diazo-trophic cyanobacteria occasionally 'bloom' in waters contaminated with phosphates and cause expensive pollution problems. No doubt yet other local risks could be foreseen, but they fade into insignificance when considered in relation to the possibility of feeding a world population of over 8×10^9 people: this is what is promised by the judicious spread of really effective nitrogen-fixing ability to cereals such as maize, wheat, barley and, more important in world terms, rice.

Further reading

Books

Dilworth, M. J. & Glenn, A. R. (eds.), 1991. *Biology and Biochemistry of Nitrogen Fixation*. Amsterdam: Elsevier.

Stacey, G., Burris, R. H. & Evans, H. J. (eds.), 1992. *Biological Nitrogen Fixation*. London: Routledge, Chapman & Hall.

Practical

Bergersen, F. J. (ed.), 1980. *Methods for Evaluating Nitrogen Fixation*. Brisbane: Wiley.

Somasegaran, P. & Hoben, H. J., 1994. *Handbook for Rhizobia: Methods in Legume-Rhizobium Technology*. Berlin Heidelberg New York: Springer.

Symposia

Bergersen, F. J. & Postgate, J. R. (eds.), 1987. A century of nitrogen fixation research: present status and future prospects. In *Proceedings of the Royal Society of London*, series B, **317**: 65–297. (This included P. S. Nutman's Centenary Lecture on the history of the subject.)

Elmerich, C., Kondorosi, A. & Newton, W. E. (eds.), 1998. *Nitrogen Fixation for the 21st Century*. Eleventh International Congress on Nitrogen Fixation, Paris. Dordrecht: Kluwer.

Tikhonivich, I. A., Provorov, N. A., Romanov, V. I. & Newton, W. E. (eds.), 1995. *Nitrogen Fixation: Fundamentals and Applications*. Tenth International Congress on Nitrogen Fixation, St Petersburg. Dordrecht: Kluwer.

Reviews

Benson, D. R. & Silvester, W. B., 1993. Biology of *Frankia* strains, actinomycete symbionts of actinorhizal plants. *Microbiological Reviews* **57**: 293–319.

Fay, P., 1992. Oxygen relations of nitrogen fixation in Cyanobacteria. *Microbiological Reviews* **56**: 340–373.

Fischer, H. M., 1994. Genetic regulation of nitrogen fixation in rhizobia. *Micro-

biological Reviews **58**: 351–386.

Hill, S., 1988. How is nitrogenase regulated by oxygen? *FEMS Microbiology Reviews* **54**: 111–130.

Howard, J. B. & Rees, D. C., 1994. Nitrogenase: a nucleotide-dependent molecular switch. *Annual Reviews of Biochemistry* **63**: 235–264.

Merrick, M. J. & Edwards, R. A., 1995. Nitrogen control in bacteria. *Microbiological Reviews* **59**: 604–622.

Peters, J. W., Fisher, K. & Dean, D. R., 1995. Nitrogenase structure and function: a biochemical-genetic perspective. *Annual Reviews of Microbiology* **49**: 335–366.

Postgate, J. R., 1989. Trends and perspectives in nitrogen fixation research. *Advances in Microbial Physiology* **30**: 1–22.

Various authors review the enzymology of nitrogenases, 1996. *Chemical Reviews* **96**: 2965–3030.

Global problems

Kennedy, I. R. & Cocking, E. C., 1997. *Biological Nitrogen Fixation: the Global Challenge & Future Needs.* University of Sydney: SUNFix Press.

Index